Equine Stud Management

Equine Stud Management

Melanie Bailey

J. A. Allen

London

© Melanie Bailey 2003

First published in Great Britain 2003

ISBN 0 85131 836 3

J. A. Allen
Clerkenwell House
Clerkenwell Green
London EC1R 0HT

J. A. Allen is an imprint of Robert Hale Ltd

The right of Melanie Bailey to be identified as author of this work has been asserted in accordance with the Copyright, Designs and Patents Act 1988

British Library Cataloguing in Publication Data
A catalogue record for this book is available from the British Library

Typesetting from author's disk, and production editing: Bill Ireson
Artwork: Maggie Raynor
Cover design: Nancy Lawrence
Colour separation by Tenon & Polert Colour Scanning Ltd, Hong Kong
Printed in Singapore by Kyodo Printing Co (S'pore) Pte Ltd, Singapore

Contents

Foreword by Joe Grimwade, Stud Manager to
    Her Majesty The Queen, The Royal Studs   ix

Acknowledgements   xi

1   The Mare   1
    Anatomy of the Reproductive System
    The Oestrous Cycle
    Factors Affecting the Oestrous Cycle

2   The Stallion   15
    Anatomy of the Reproductive System
    Physiology
    Semen Production
    Factors Affecting Semen Production
    Assessing Fertility

3   Breeding   26
    Choosing the Mare and Stallion
    Teasing

The Problem Mare

Preventing Infection

Breeding Methods

Mating

4   Pregnancy   67

Foetal Development

The Hormones of Pregnancy

Pregnancy Diagnosis

Care of the Pregnant Mare

Problems Affecting Pregnancy

5   Abortion   86

Non-infectious Causes of Abortion

Infectious Causes of Abortion

Isolation Procedures

6   Foaling and the Newborn Foal   102

Monitoring Foaling Mares

Foaling Equipment

Preparation for Foaling

Normal Foaling

Foaling Problems

Problems Affecting the Newborn Foal

Orphan Foals and Fostering

7   The Foal from Birth to Weaning   143

Early Handling

Exercise and Behaviour

Nutrition

Teeth

Routine Care

Weaning

8   Stud Design and Administration   158

Layout and Facilities

The Breeding Area

The Foaling Box

Paddocks and Fencing

Fire Safety and Emergency Procedures

Studfarm Records

Index   **185**

Foreword

Having known Melanie for over fourteen years, it comes as no surprise to find that she has obvious talents as an author!

Our first contact was when I was the Assistant to the Manager at The National Stud where Melanie was attending the Student Training Course. The course aimed to develop a sound theoretical knowledge combined with the practical skills associated with working on a stud-farm and Melanie excelled in every aspect; she left The National Stud, as she had already left Warwickshire College, with the highest accolades.

Having already gained experience through the Pony Club as well as working in Polo, Dressage and National Hunt yards, Melanie then set about expanding her excellent Curriculum Vitae even further. She gained experience of the very different demands of stud work in the Southern Hemisphere and then returned to Britain to make the difficult transition from working in the yard to working in the office, both as as secretary and as my assistant with particular responsibility for the student training course that she had previously attended.

Now that the time has arrived to dedicate much of her time to the demands of wife and mother, she has turned her hand to writing. The result is a wonderful opportunity to share some of the knowledge that Melanie has acquired through her training and experiences. The book succeeds in providing a solid introduction to the most important elements of breeding horses and ponies with sections on the mare, the stallion, the stables, the paddock and the office. This in itself is an admirable achievement, but it also manages to provide the information in such a way that it does not exclude any discipline or level of participation in horse and pony breeding worlds.

For the person who is considering breeding a foal, it can provide much of the knowledge that is, regrettably for both horse and owner, often learnt by bitter experience. Equally, for students preparing to embark on careers in the breeding industry, it provides a short cut to knowledge that it can take years to acquire. Melanie has shown that she possesses a wonderful balance of knowledge and experience as well as the ability to provide information at a consistent level in an easy to read text. Whatever route the reader is following, this will be an excellent starting point as well as a valuable point of reference for the future.

Joe Grimwade
Stud Manager to Her Majesty The Queen, The Royal Studs

Acknowledgements

To Dr Peter Rossdale MA PhD DESM FRCVS OBE, I would like to extend my deepest appreciation for all his support, advice and enthusiasm over the years, which has made writing a book like this possible, as well as for supplying the majority of the photographs used in this book.

I would also like to offer special thanks to Joe Grimwade, Stud Manager to Her Majesty The Queen, for providing the foreword for this book but most importantly for his invaluable friendship.

A heart felt thank you also to Brigitte Heard of Rossdale and Partners, who makes the seemingly impossible occur like a routine daily event.

To Professor 'Twink' Allen of the Equine Fertility Unit, Newmarket for giving his valuable time and providing the information regarding equine identical twins.

Last but not least, to Caroline Burt, Editor, J A Allen, for her exceptional patience and understanding.

1 | The Mare

The mare's reproductive system consists of two ovaries and fallopian tubes, a uterus, vagina, clitoris and vulva. All of these structures are closely grouped, however there are other important parts of the system – the pineal gland, hypothalamus and pituitary gland, which are located close to the base of the brain. The latter structures produce some of the main sex hormones, which are the chemical messages sent from one part of the body to another. The oestrous cycle is discussed in detail later in this chapter.

Anatomy of the Reproductive System

The genital tract of the mare is designed to provide the optimal environment for reproduction to occur. The tract can be divided into internal and external structures (see Figures 1.1 and 1.2).

External Anatomy

Vulva and Clitoris

Both lie immediately beneath the anus, under the tail. The vulva is composed of two lips, or labia, and it acts to form an airtight 'door' or

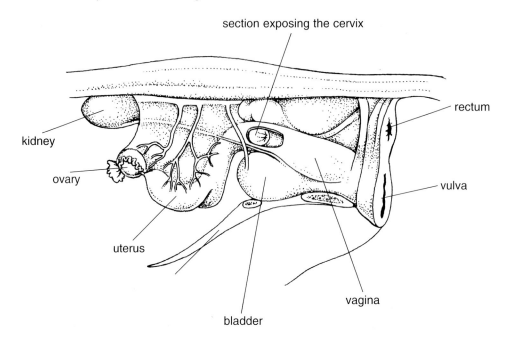

section exposing the cervix

rectum

kidney

ovary

vulva

uterus

vagina

bladder

Figure 1.1 *The reproductive organs of the mare*

seal to the inner structures of the genital tract. At the base of the vulva is the clitoris. The clitoris lies inside a little pouch and has three small cavities, called sinuses. These cavities contain a varying quantity of a 'cheesy' or 'waxy' substance known as smegma, which is also found on the stallion's penis and sheath. The clitoral sinuses are significant as a site for possible venereal (contracted sexually) infection. (Routine testing for infection is discussed more fully in Chapter 3.) The clitoris is normally hidden from view, but the mare is able to pull back the vulval lips to expose the clitoris which she does when she is in oestrus or 'in season' as an important part of her sexual behaviour and response towards the stallion. (Sexual behaviour during oestrus is discussed more fully later in this chapter and in Chapter 3.)

The conformation, or shape, of the vulva is particularly important. In older mares, the vulva tends to become progressively more flaccid and sloping. In fact, some mares may naturally have this type of vulval conformation from an early age. Due to the position of the vulva just below the anus, poor conformation results in faeces being deposited directly onto the labia and thereby contamination of the tract can occur. Poor vulval conformation also means that the important seal formed by the labia is weak and air can be readily taken into the genital tract. Almost all air is rich in environmental pathogens (disease causing organisms) that thrive in the dark, moist conditions of the internal genital tract. Mares that have poor vulval conformation may be difficult to breed from successfully and management of these mares is discussed more fully in Chapter 3.

Internal Anatomy

Vagina and Vestibule

The vagina is a tube-like passageway between the womb and the vulva. It is normally divided into two sections, anterior and posterior. The posterior part is just inside the vulval lips and the anterior part near to the opening to the womb. The vagina is designed to provide a suitable place for the stallion's penis during mating, allowing semen to be deposited directly into the womb, or uterus, as well as serving as a birth canal for the foal at the end of pregnancy. It is capable of extreme dilation due to its muscular walls and has glands which secrete lubrication during oestrus to aid the passage of the stallion's penis during mating.

In maiden, or unmated mares, a thin membrane called the hymen may be present although it is not always present as a complete membrane. The vestibule is the name given to the posterior part of the vagina, just inside the vulval lips. The vestibule also provides for the opening of the urethra, the tube leading from the bladder, to allow passage of urine. In the normal mare, the floor of the vestibule is positioned so that urine cannot flow backwards into the vagina.

Cervix

A muscular constriction between the vagina and uterus, the cervix is often described as having the appearance of a rosebud. It protrudes into the vagina and has a strong muscular seal to protect the uterus from any outside infection, giving it a vital role during pregnancy.

The muscular seal of the cervix is affected by the hormone progesterone (the reproductive hormones are discussed in detail later in this chapter). Progesterone causes the cervix to remain tightly closed during pregnancy and when the mare is not in season. When the mare is in season, the cervix relaxes to allow the stallion's semen to be deposited directly into the neck of the uterus during mating. When the mare is giving birth the cervix dilates to its fullest extent to allow the passage of the foal.

Uterus

The hollow, muscular structure that provides a protective, nourishing environment for the foetus, the uterus is an extraordinary organ. It undergoes enormous changes in size and structure throughout the eleven month pregnancy and yet is able to return to its non-pregnant size and shape very quickly after the foal is born.

The uterus is termed as being 'Y' shaped (Figure 1.2). It has a body and two horns. The body of the uterus connects with the vagina via the cervix. The uterus is made up of three main layers. The structure of these layers changes during pregnancy to provide optimum protection and nourishment for the growing foal. The first layer is a protective outer coating which surrounds the whole of the uterus; below this, the second layer contains several strata of muscle, blood vessels and nerve cells; the third or inner layer is called the endometrium and contains glands and connective tissue. The endometrium provides the early nourishment for the foetus and it is to this surface that the placenta will attach. If this layer of the uterus is damaged by infection or age, the mare will probably fail to conceive or to carry a healthy pregnancy to full term (see Chapter 3). Unlike the human female, the mare does not shed the lining of the uterus on a regular basis throughout her reproductive life so any damage to the endometrial layer will invariably be permanent.

Broad Ligaments

The uterus is suspended in the abdomen by broad ligaments designed

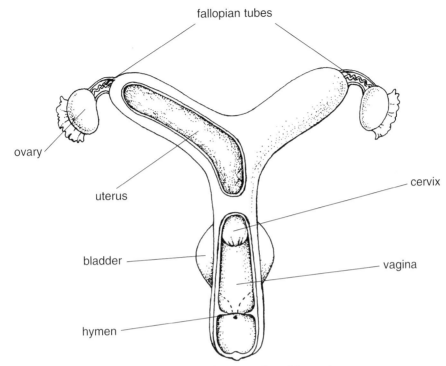

Figure 1.2 *The reproductive organs of the mare viewed from above*

to support the uterus and ovaries. Each of these large ligaments contains blood vessels and nerve cells which supply both the uterus and ovaries.

The ligaments suspend the uterus in such a way as to assist in correct drainage of fluids from the genital tract, as well as providing vital support for the uterus throughout pregnancy. As the mare gets older, or after many pregnancies, these ligaments can become stretched and predispose her to infections due to poor drainage.

Fallopian Tubes

At the end of each uterine horn there is a fallopian tube. Each of these long coiled tubes ends in a fringed structure known as the infundibulum. This structure partially overlaps the part of the ovary where the egg will be released during ovulation; the fringe-like end is designed to guide the egg into the fallopian tube. Each fallopian tube is about 20–30 cm long when uncoiled.

When the mare is mated the stallion's sperm will swim to the fallopian tube and fertilize the waiting egg. The egg must be fertilized at the right time and only fertilized eggs will continue the journey down into the uterus itself. This journey from tube to uterus takes the fertilized egg about 5–8 days. Each of the fallopian tubes has an inner lining of microscopic finger-like protections, called cilia, which aid the movement of the sperm to the egg and of the fertilized egg into the uterus.

Ovaries

The mare has two kidney-shaped ovaries, one at the end of each of the fallopian tubes. Each ovary is held within the broad ligament and each is covered with a tough layer of tissue which covers all but a small area on each ovary – the uncovered area is located in the 'indent' of the kidney shape and it is through this site that ovulation occurs. In many other species, ovulation can occur anywhere on the surface of the ovary, but in the horse the tough tissue layer ensures that it is only possible at one site.

Each ovary contains many eggs, or ovum, which will be released, one or two each heat period, into the fallopian tube ready for fertilization if the mare is mated. Each female horse is born with eggs already present in her ovaries and she will not produce any more during her lifetime. This is one of the reasons why she has two ovaries, so that there is a 'back up' should one fail – mares can still be fully fertile with only one normal ovary. In simple terms, the egg is prepared for ovulation within a blister-like sac on the ovary called a follicle. When the mare is in season, hormones produced by the ovary itself cause one or two of the follicles to increase in size. (The hormones involved in ovulation are discussed in detail later in this chapter.)

Although there may be many follicles present in each ovary, normally only one or two will reach the size and condition required for ovulation – that is, about 4–5 cm in size. By means of rectal palpation, a procedure carried out by a veterinary surgeon when examining the mare, these large follicles can be felt or scanned through the rectum wall.

The Pineal Gland, Hypothalamus and Pituitary Gland

The pituitary is a very small gland situated at the base of the brain. It is stimulated by two other structures also located close to the brain – the hypothalamus and the pineal gland. The pineal gland responds to the level of light falling on the back of the eye, in fact to the number of hours of daylight. The hypothalamus is stimulated by the pineal gland and it in turn, stimulates the pituitary gland to release certain sex hormones which target the reproductive organs. The pituitary gland produces a wide range of hormones, conveying chemical signals to a number of organs via the bloodstream. These hormones are so effective that extremely small quantities can produce dramatic effects on the functioning of the whole body, the target organ and/or on the behaviour of the horse. The effect each of the sex hormones has in coordinating the reproductive cycle of the mare is very important, in fact without the correct levels of each she would not be able to breed at all.

The Oestrous Cycle

The oestrous cycle is the term used to describe the breeding cycle of the mare throughout the year. The way the cycle works can appear to be very complicated, but understanding the basics of each stage is vital for all breeders or students of equine reproduction.

In simple terms, the cycle can be broken down into two main parts. Oestrus is the stage when the mare is receptive to the stallion and when her body is prepared for the optimum chance of conception should she be mated. It normally lasts for 4–6 days, but can be as short as two days in some mares. Dioestrus is the stage when the mare is not receptive to the stallion and during which she can be extremely aggressive towards him. It normally lasts for 14–16 days. The normal mare will continue to cycle, oestrus then dioestrus, during the breeding season until she is pregnant when she will remain in a dioestrus-like state.

The mare is termed a seasonally polyoestrous breeder and has a gestation period, or pregnancy of eleven months. Seasonally polyoestrous simply means that the mare has several breeding heats at certain times of the year. Nature has designed the mare to breed only at those times of the year when there is an optimum chance of reproductive success – during the spring and summer months. Therefore, a foal conceived in May will be born the following April when, in the northern hemisphere, there is a flush of spring grass and the weather is usually improving. During the late autumn and winter, many mares stop cycling all together and their reproductive systems essentially become dormant – a stage known as anoestrus.

In the late spring, when daylight length increases, the mare's reproductive system is stimulated into action. The amount of light falling onto the back of the eye stimulates the pineal body to, in turn, stimulate the hypothalamus to produce gonadatrophin-releasing hormone (GnRH). (Gonad is the generic name given to the sex organs: the ovaries in the mare and the testes in the stallion.) GnRH is released in the bloodstream and its target organ is the pituitary gland. The pituitary gland releases follicle-stimulating hormone (FSH), which targets the ovaries and causes follicles to develop. The ovarian follicles themselves act as glands and produce the powerful female hormone, oestrogen. Oestrogen causes dramatic behavioural changes in the mare, altering her violent refusal of the stallion to active acceptance. Oestrogen also causes changes within the genital tract to improve the environment for mating. The vulval lips become engorged and relaxed in appearance, the vagina becomes flushed with blood and lubricating fluids are produced. The surge of oestrogen circulating in the bloodstream stimulates the pituitary gland to now produce luteinising hormone (LH) which causes one of the follicles developing on the ovary to mature and ovulate, releasing the egg into the fallopian tube. The act of ovulation causes the blister-like follicle to burst and leaves a cavity within the ovary. The cavity rapidly fills with blood and is called the *corpus luteum* (CL) or yellow body. The yellow body produces the hormone progesterone, often called the hormone of pregnancy.

Progesterone is the dominant hormone during dioestrus when the mare will not accept the stallion. If the mare is not mated, or the mating is unsuccessful, the yellow body will continue to function for about 14 days, when prostaglandin, another hormone, is released by the uterus. Prostaglandin causes the yellow body to stop functioning and for the cycle to begin again with the release of FSH from the pituitary.

Factors Affecting the Oestrous Cycle

There are many reasons why the mare may have problems during the breeding season. In some breeds, where an artificial breeding season or method is desired, these problems can occur quite frequently.

Management

The most important factor affecting the oestrous cycle is the way the mare is managed. An important point to remember is that the horse has evolved from a small dog-sized mammal, walking on three toes to the large, single hoofed mammal of today. The rich variety of type amongst the horse breeds has mainly been brought about by human intervention, adapting and developing certain traits to produce specific breed types for specific purpose, such as the Shire for heavy draught work and the Thoroughbred, the ultimate equine athlete. Although the breed types have been developed by man, the basic needs of each horse remain the same as they have been for many thousands of years; this fact alone accounts for many of the problems encountered with the management of horses in a domestic environment. Almost every aspect of the way we keep our horses is artificial and very unlike the way the horse would live in the wild. Thankfully, horses are wonderfully adaptable and most remain well and healthy in this artificial state.

Controlled Breeding Seasons

The imposition of set breeding seasons for breeds such as the Thoroughbred is one of the first factors affecting the mare's normal oestrous cycle. As already discussed, the mare would normally begin to cycle in late spring through to early autumn. The Thoroughbred season is set to begin on 15 February, a time when the mare may still be in deep winter anoestrus, or changing from anoestrus to regular cycling. This stage is called vernal transition and it usually occurs in late winter or spring, depending on the individual mare and how she is looked after. During this time, she may show very erratic heat patterns, she may remain in oestrus for a long time and then fail to ovulate before returning to dioestrus. This stage can last for some weeks and can cause considerable frustration for the breeder who wishes to get his or her mare in foal early in the season.

One way to solve this problem is to fool the mare's reproductive system into thinking that spring has come early. We know that the main trigger for cycling to begin is the amount of light falling onto the eye and normally, early in the year, the amount of natural daylight is still limited. However, by exposing the mare to artificial lighting, simply by leaving the light on in her stable in the evenings and early mornings, it is quite possible to encourage her to begin cycling even in the depths of winter. The use of lighting in this manner has long been a standard procedure on many studfarms, particularly those imposing an early breeding season. Although a lot of research has been carried out within the last few years on the actual duration and strength of light needed, the most common method is to use a minimum of a 100 watt bulb and to leave the lights on each evening from dusk until about 10.00 p.m. Some studfarms have a timer system to automatically turn the lights on and off at the correct time and to allow for an additional lighting period to begin in the morning at about 4.00 a.m. This system works just as well by extending the evening light only and may be more practical for the small breeder who does not have, or cannot justify financially, an automated lighting system.

The main disadvantage with this latter system is that, for optimum success, it must be started as early as possible – most studfarms begin additional lighting from the November prior to the breeding season. Some studfarms gradually increase the amount of light the mare is exposed to over the following weeks to further mimic the gradual increase in daylight; however, recent studies have indicated that this is not necessary. The other main disadvantage is that although many mares remain living out until late December, the mare must be stabled for the lighting system to work and this will obviously mean additional keep costs for the breeder as well as additional labour costs. Stabling the mare by night will also allow her to be kept in a warmer environment than that of living out and some breeders also like to rug their broodmares to encourage early cycling. However, at present there is no known link between external temperature and early cycling.

Physical Condition

The physical condition of the mare can also affect her ability to cycle normally during the breeding season. A mare that is in poor condition through poor nutrition or illness is unlikely to have the optimum chance of conceiving early in the year. Ideally, mares should be on a rising plane of nutrition going into the breeding season to simulate the flush of spring grass that would be available to the mare in the wild state. Conversely, allowing the mare to get too fat may also affect her ability to cycle; in fact, many obese mares have erratic cycles.

Physical Fitness

Mares going to stud for the first time, especially those that are at a high level of fitness such as racehorses and performance horses, may need some considerable time to begin cycling properly. Often, athletic mares competing at high levels cease cycling altogether and require a considerable 'let down' period of some months to establish normal oestrus activity. 'Letting down' is the term used when a horse is turned out to grass and/or given a holiday from training and competing. In many cases, the horse responds well to a complete break and his/her physical shape changes as the muscle tone softens – hence the term 'letting down'. Ideally, mares that are to be retired to stud should begin their preparation for breeding five or six months prior to the start of the season. In the northern hemisphere, this would mean making the gradual change from full work to no work from late summer/early autumn. Obviously, this is not always possible and some mares are sent to stud late in the season due to an injury that means they have to take some time away from full work. Additionally, mares may be also sent to stud whilst still in full work so that they can be mated and return to work for the early part of their pregnancy. This happens commonly with performance and race mares, where the mare's ability is such that she will be bred from at the pinnacle of her career, when the value of her progeny will be perhaps at its greatest. Although prohibited in Thoroughbreds, the use of artificial insemination and embryo transfer in the performance breeds means that an accomplished mare need not retire from competing to have a foal. (These techniques are discussed more fully in Chapter 3.) The fact that the mare's own fitness may cause a problem with her breeding plan needs to be taken into consideration before she is sent to stud.

Physiological Factors

As already discussed, the reproductive cycle of the mare is controlled by hormones. The hormones can be likened to chemical messengers that are released from a gland and travel through the bloodstream or lymphatic system to a target organ.

The actual amount of hormone needed to trigger a reaction is very, very small – nanograms per millilitre of blood to be precise – which is one thousand millionth of a gram per millilitre of blood. Even in such minute quantities, these hormones can cause rapid and dramatic changes. The system relies on an intricate interrelationship between each of the hormones, one triggering the release of another. In a recent work, *The Horse from Conception to Maturity* (J. A. Allen 2002), Peter Rossdale has used an analogy of locks and keys to describe this inter-relationship. Dr Rossdale explains that the hormone itself may be likened to a key produced by the gland which then, in turn, fits into the lock of the target organ; for example, the pituitary gland produces the keys of FSH which fit into the locks on the lining of the follicles of the ovary. However, no matter how many keys are produced by the gland, they will not work unless the same number of locks are present in the target organ. Therefore, there will be times when there are more locks than keys, or conversely, more keys than locks. All hormones have the ability to change the number of locks of the target organ, also known as receptors, but at times, such as when the mare is beginning to cycle again after her winter break, the locks may not be sufficient for the keys or vice versa.

Obviously, these factors can cause problems with the normal oestrous cycling of a mare and, apart from erratic cycling, there are other relatively common physiological problems. Some of these are:

- Silent heat – when the mare comes into season and ovulates without showing the behavioural signs of oestrus.
- Persistent *corpus luteum* – when the mare remains in dioestrus, even though she is not pregnant, long past the normal 15–17 days.
- Lactation anoestrus – when the mare fails to return to oestrus after giving birth to her foal, or after the first 'foal' heat.
- Prolonged oestrus – when the mare remains in oestrus for a long period.

These are just a few examples of the problems that can be encountered when the mare's reproductive system is not functioning correctly. Most are short-lived and reversible, causing no lasting problem other than loss of time during the breeding season. Today, on most modern studfarms, the veterinary surgeon has become very much part of the team, visiting the farm regularly during each week of the breeding season, sometimes every day. In addition, the knowledge and expertise of most stud staff has also improved in recent years due to the variety of training courses available. Both of these factors have resulted in a greater understanding of the physiology of broodmares and quick identification of an individual who may have a problem. In the last 50 years, equine reproduction techniques have changed beyond recognition and veterinary surgeons now have a vast array of diagnostic tools and treatments available to them. The use of synthetic hormones to assist with cycling problems or to improve fertility rates in horses has become relatively commonplace, again the technical details of these treatments are beyond the scope of this text but some further aspects of breeding from the problem mare are discussed in Chapter 3.

2 | The Stallion

The stallion is the figurehead of the studfarm and most stallions have a very high financial value, none more so than the Thoroughbred. Stallions are considered more valuable than mares due to the fact that they can produce many offspring each year, whereas, with natural methods, the mare will only normally produce one.

Keeping stallions on a studfarm requires special facilities in almost all cases to ensure the health and safety of both the horses and their handlers. Stallions, whatever their breed, do require experienced handlers and carefully planned procedures – the potential dangers of any stallion should not be underestimated as it is usually through complacent handling that accidents occur.

The health and well-being of the stallion is paramount, as his fertility status will directly affect the studfarm's success and profits. For the breeder, whatever breeding method is used, a basic understanding of the stallion's reproductive system is essential to ensure optimum fertility rates are achieved.

Anatomy of the Reproductive System

The stallion's reproductive anatomy consists of organs that produce, store and nourish the spermatozoa as well as the penis, the organ that

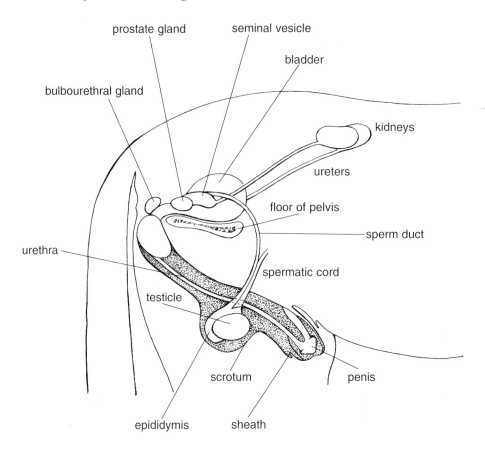

Figure 2.1 *The reproductive organs of the stallion*

deposits the semen into the mare's reproductive tract (Figure 2.1). In addition, there are glands that produce seminal fluids and the reproductive hormones.

Scrotum

The stallion's scrotum is located behind the penis. A fleshy bag, or sac, it contains the stallion's testes. When the stallion mounts the mare, care should be taken to protect the scrotum from injury.

Testes

In the normal stallion, there are two testes, or testicles, which are located on the abdomen, just forward of the hind legs. These are the organs that produce the sperm and produce the male reproductive hormone, testosterone.

The testicles are enclosed in the scrotum and each is suspended from the abdomen by connective tissue and muscle which forms the spermatic cord. This explains how the stallion is able to draw up the testicles in cold weather, or when they may be at risk of injury – such as when jumping. The spermatic cord also supplies each testicle with blood vessels and nerve cells. External temperature is an important factor in sperm production, as described later in this chapter and in Chapter 3; the viability of the sperm and the length of time it is active is greatly affected by temperature. Body temperature is too hot for optimum sperm production and this is why the testicles hang outside the body; however, should the external temperature drop, the testicles can be drawn back up into the abdomen. The need to constantly regulate temperature is also the reason why the scrotum is normally hairless in the horse.

During the colt foal's development in the womb, the testicles are located high up in the abdomen, close to the kidneys. Shortly before birth, each testicle migrates down to the scrotum, via a passageway called the inguinal canal and through a hole in the membrane surrounding the abdominal cavity, called the inguinal ring. If the hole in the membrane is too large, there is a risk that a portion of the stallion's intestines could pass through resulting in an inguinal hernia. This condition can be serious and should always be assessed by a veterinary surgeon. Sometimes, one or both of the testicles are retained in the abdominal cavity, either temporarily or permanently, and later on this can cause fertility problems to the stallion. Horses affected in this way are known as cryptorchids or 'rigs'. As sperm production is affected by temperature, the stallion's own body temperature inside the abdomen will stop the production of sperm in the affected testicle, or testicles. Normally, stallions with both testicles retained are infertile and those that have only one testicle affected are fertile. There is some evidence that this condition is inherited and, ideally, an affected horse should not be used for breeding.

Epididymis

A long tube that tightly coils onto itself, the epididymis is attached to the testicle and is where the sperm mature when they have left the testicle.

Spermatic Cord

The cord contains the muscles, nerves and blood vessels that serve the testicle. It also contains the structure known as the *ductus deferens* or *vas deferens*.

Vas Deferens

A muscular tube, which is a continuation of the epididymis. Its function is to transport sperm from the epididymis to the urethra ready for ejaculation. In the abdomen the *vas deferens* becomes dilated to form a temporary storage area for mature sperm, this 'store' is called the ampulla.

Urethra

The tube that connects the bladder to the penis, the urethra allows for the passage of urine as well as transporting semen during ejaculation. At one section of the tube it is joined by three glands, jointly called the accessory glands.

Accessory Glands

These consist of three glands: the prostate gland, the bulbourethral gland and the seminal vesicles. Their role is to produce seminal plasma, the actual function of which is not fully understood. It may be that the seminal plasma acts as transport fluid during passage into the mare's uterus as well as a temporary energy source for the sperm. Some stallions also produce a gel-like substance from the seminal vesicles at the end of ejaculation, which can be evident when the stallion dismounts, at the mare's vulva or on the stallion's penis.

Penis

The penis is an organ designed to allow safe passage of the sperm into the mare's genital tract. It is made up of a special muscle type that allows blood to fill the penis during sexual arousal. The blood is held within the tissue and cannot escape normally until after ejaculation. The penis doubles in length and thickness during erection and the head, called the glans, trebles in size. The glans increases so much in size in order to dilate the mare's cervix during mating, ensuring that the semen is deposited into the neck of the uterus upon ejaculation.

At the end of the penis, the urethra can be seen protruding a few millimetres beyond the glans. Around the urethral opening is an area known as the urethral fossa, similar to that described earlier in relation to the clitoris of the mare. In this fossa an accumulation of smegma occurs which may provide an ideal environment for venereal diseases; because of this, the tip of the penis is routinely swabbed. (Routine stallion swabs are discussed in more detail in Chapter 3).

Sheath

The sheath, or prepuce, is the fold of skin that houses the penis and continues from the scrotum, which houses the testicles. At rest, the sheath is folded back on itself with the penis withdrawn inside. The folds of the sheath also produce smegma and this is another area that is routinely swabbed for breeding stallions.

Physiology

We have already examined the breeding cycle of the mare in some detail and the stallion's cycle is quite similar in many ways, although he does not have to rely on periods of oestrus for mating to be possible. The stallion is a seasonal breeder; in his natural state there would be little point in him wanting to mate with his mares in the winter months when they were not cycling at all. Although the stallion continues to produce sperm all year round, production is less prolific in the winter; it is possible for him to mate during the winter months, but it may take him considerably longer to become sufficiently aroused to reach ejaculation and the quality of his sperm may be poor.

The stallion's reproductive system is also stimulated by the increase in daylight length in springtime and, in response to this stimuli, the hypothalamus produces GnRH, just as in the mare. This hormone targets the pituitary to produce FSH and LH, again just as in the mare. However, in the stallion, FSH causes the growth and formation of sperm in the testicles and LH causes special cells in the testicles, the interstitial cells, to release testosterone into the bloodstream. Testosterone is the male hormone that causes masculine behaviour and libido (sexual drive).

Semen Production

Sperm (spermatozoa) are the male sex cell, just as the ovum or egg is the female sex cell. Each sperm has a head – which contains the cell nucleus, a middle piece and a tail (Figure 2.2) and is about the size of a grain of sand. Each normal sperm is capable of swimming a considerable distance to meet the waiting egg in the mare's fallopian tube.

Sperm are produced in the testicles and during this process the number of chromosomes contained in the cell nucleus is reduced by half to 32. When fertilization occurs, the egg supplies the other half of the chromosomes to make up the normal number of 64. In a normal ejaculate, the stallion releases an enormous number of sperm, about 100 to 150 million per cubic millimetre and yet only one sperm will fertilize the egg. The reason for this is thought to be that the sperm have a cumulative effect on the egg to allow fertilization to occur, but the precise number necessary is not yet known. However, it is known

end piece   tail   mid piece   head

Figure 2.2 *A normal sperm*

that a layer of debris surrounds the egg and the sperm work to remove this before one achieves fertilization, so it is possible that the high number of sperm is necessary for this action to be successful. Additionally, sperm must also go through a final period of maturing in the fallopian tube before they can fertilize the egg and, again, it may be that a large number is needed to ensure optimum success. As we can now see, sperm production is a complicated, somewhat lengthy process – it actually takes about six weeks to produce the mature sperm ready for ejaculation during mating.

Sperm are ejaculated together with seminal plasma, but once in the uterus it is thought that the sperm separate from the plasma. The sperm are moved by contractions of the mare's uterus and actively swim to the openings of the fallopian tubes. As mentioned earlier, the function of plasma is not fully understood, but it has been suggested that apart from assisting with transport of the sperm to the uterus and providing some form of nourishment, plasma might also kill off any sperm left behind in the uterus to avoid the egg being fertilized with an 'old' or weak sperm. Sperm age rapidly following ejaculation and are very sensitive to the condition of the uterus (see Chapter 3), but normally the sperm will remain fertile for about two days following ejaculation. How long each stallion's sperm lives is subject to huge variation and each stallion is different; some individual stallion's sperm is reported to have caused fertilization seven days after mating. Taking a sample at regular times throughout the season is one way that large studfarms assess their own stallions and, for those farms that routinely use artificial insemination techniques, this can be done very easily. Sampling also allows each stallion's fertility to be assessed if he has been particularly busy, or if he has had problems such as a number of his mares not getting in foal.

Factors Affecting Semen Production

Sperm are produced throughout the stallion's reproductive life, unlike the mare who has all the eggs she will ever produce present in her ovaries at birth. However, although there is constant production, sperm are very sensitive to changes in their environment, whether it occurs when they are still inside the stallion or not.

Infertility is a term that is often used to describe a stallion, or a mare, that has a period of difficulty in reproducing. However, infertility is not strictly speaking the correct term for any horse that has bred successfully in the past. Complete infertility or sterility is uncommon and normally is due to an inherited problem, something the horse was born with. A horse can become infertile due to injury or illness, but a temporary period of being unable to breed is normally defined as sub-fertility.

Illness

If the stallion is ill and has a raised temperature, this will affect his sperm production. With today's modern medications, it is unlikely that a stallion will run a temperature for more than a few days, but a prolonged period of serious illness can cause some degree of sub-fertility which is usually temporary, taking about two months for the sperm production to return to normal. In some cases however it may result in permanent sub-fertility. Any illness in a breeding stallion is therefore of concern, especially when that stallion may also be subject to a restricted breeding season and time is limited.

Injury

Injury to the genital organs is quite common in stallions that are bred in-hand (see Chapter 3). These injuries usually take the form of kicks from the mare during the time that the stallion is mounting or dismounting. Any resulting wounds, swelling and/or bruising can – by increasing the circulation to the area and therefore also increasing the temperature – affect sperm production. It is imperative that any injury to the genital organs should be assessed by a veterinary surgeon as soon as possible, to reduce the risk of permanent damage occurring. Wounds to the penis can be difficult to treat as it is normal for the penis to withdraw into the sheath at rest; if, due to inflammation, the penis cannot withdraw, then long-term problems can result, particularly if the nerve and blood supply to the organ are also affected. (Safe practice in the breeding shed is discussed in Chapter 3.)

Drugs

Today it is uncommon for stallions to be given any drugs without due consideration for the effect they may have on sperm production. Some steroid-based treatments – particularly anabolic steroids, given more commonly when the horse is in training – can completely stop the production of sperm. However, production usually returns to normal after about three months. Some tranquillizers may also have an adverse effect on breeding stallions. This is because the drugs may cause the penis to protrude from the sheath for a long period, resulting in swelling and fluid accumulation. As with an injury to the penis, this can have serious long-term effects and in the worse case scenario may require amputation. However, as the complications of such drugs are now well known and other safe alternatives available, it is unlikely that any veterinary surgeon would administer them to a stallion.

Management

How a stallion is managed can affect his fertility and his physical ability to breed. There are many factors that can affect the domesticated stallion, but here we deal with those that specifically affect his ability to produce sperm. The two most significant management factors that may affect semen production are nutrition and over-use.

Nutrition with regard to optimum semen production is still an area that is not fully understood. It is known, however, that stallions that are in poor condition or are receiving poor quality feedstuffs, will not produce good quality semen. Some studfarms give vitamin and mineral supplements to their stallions during the breeding season, Vitamin E based additives being the most commonly used. Feed for working stallions should follow the same common sense guidelines as feeding any other horse for performance, feeding for breed/type, according to workload and avoiding any sudden diet changes, etc.

Over-use of a stallion is likely to be a more common problem on Thoroughbred studfarms where only natural conception methods are permitted. As we will see when we discuss the use of insemination as a mating method for non-Thoroughbreds in Chapter 3, splitting ejaculates is possible in most cases, allowing a number of mares to be inseminated successfully with only part of one ejaculate. Some Thoroughbred stallions are extremely popular and it is surprisingly common for a stallion to have over 100 mares to cover during the short restricted breeding season. Although modern veterinary techniques have made it possible to fine tune when mares are mated, effectively avoiding 'wasted' matings, there will still be occasions when a stallion will be required to mate three, possibly four, times in one day.

Assessing Fertility

There is a huge variation between the fertility rates of individual stallions. Some stallions' semen production remains unaffected however many times they mate during the season, others cannot cope with mating more than twice a day, every other day; each stallion must be assessed very much on an individual basis. When the stallion is retired to stud, it is rare for his semen quality to be tested prior to the start of his career. This is because a sample taken at this time can be misleading. Sperm that have been stored for a long time will age and the first few ejaculates produced will contain a large percentage of old, possibly dead, sperm. Once he begins mating regularly, newly matured sperm will be ejaculated and a sample taken at this time will present a completely different picture. It should also be pointed out that some stallions appear to have very poor semen quality on sampling, but nonetheless continue to get a good percentage of their mares in foal. Many studfarms will take random samples from their stallions throughout the season to establish what is normal for each individual and to assess his fertility rate. This is good practice as it means that potential problems can be identified early on. As sperm production takes some time, any sample will reflect the conditions of the stallion's sperm production some six weeks previously. Studfarms that use artificial insemination techniques have the advantage with regard to sampling as each collected ejaculate is routinely assessed prior to its preparation for storage or insemination.

3 | Breeding

Whether enjoyed as a hobby or conducted as a business, breeding horses, can be an expensive undertaking. For many small breeders, the first step into the stud world is when a race or competition mare retires from active performance, or perhaps when a filly turns out to have the pedigree but not the ability in her chosen field of competition.

Large or small, studfarms are only as successful as the horses they breed and careful selection of the foundation mares and stallions is vital.

Choosing the Mare and Stallion

The Mare

The choice of broodmare is important as she has the immediate maternal influence on the foal as well as passing on her genetic material and these influences are just as important as those passed on by the stallion. Some breeders will spend many hours poring over catalogues and statistics before choosing their stallion, whilst not giving very much thought to the actual suitability of that stallion for their mare.

Broodmares are the financial backbone of any studfarm and they should be chosen for their ability to produce marketable and desirable quality youngstock. Most breeders will sell on the majority of their youngstock, perhaps keeping some to compete in their chosen sphere, with the best mares coming back into the broodmare herd when their competing days are over. Other breeders may like to buy in a proven mare to complement the mares they already have, or to establish a small breeding herd. The principles behind choosing a mare remain the same, whether she is to be the only broodmare or one of many.

It should almost go without saying that the mare should be fit and healthy, although some retire to stud after suffering a training injury; not all of these mares will prove fit enough to carry a foal to term and look after it properly. Any mare with a genetic condition should not be chosen as it is likely that she will pass this condition on to her offspring. Breeding from elderly mares can also prove to be costly as the older maiden mare, a mare going to stud for the first time, may be past her breeding prime and take a long time to conceive, fail to carry her foal to term or produce poor quality foals. Quite often older proven mares come up for sale and it can be tempting to purchase such a mare who may have an excellent pedigree and have produced good offspring in the past. Before considering purchasing such a mare, her recent breeding history should be examined in detail, as it may be that she has had difficulty conceiving and carrying a healthy foal to term in recent years. A mare who has been barren (failed to conceive) for more than one year may only have a 25–50 per cent chance of getting in foal; add on to that her advancing years and the chances of her conceiving become even less likely. Any information regarding the mare's breeding history, especially her normal oestrus behaviour and perhaps even previous oestrus patterns, as well as information regarding any foals she has had is of considerable assistance. Many broodmares, especially Thoroughbreds, spend a lot of time away at stud, perhaps a different one each year and management records from the studs they have visited may be the only first hand information of their breeding patterns available.

In an ideal world, all potential broodmares should be chosen not only for their competitive ability and/or pedigree, but also for their soundness, fertility, conformation and temperament. All too often mares and stallions are chosen just on the basis of ability or pedigree alone and, ultimately, this is damaging to the breed as any unsoundness and weakness will be continually bred back into the next generation. Assessing mares prior to selection for breeding is one way where obvious problems can be identified even before any examination by a veterinary surgeon. Any checks for breeding soundness are best carried out a few months prior to the start of the season so that minor problems can be corrected in time.

The first part of any examination should be just looking at the mare and checking her physical condition and external reproductive conformation. Examining the external reproductive conformation can be done by simply lifting the mare's tail. From what we have already learned in Chapter 1, the conformation, or shape, of the mare's vulva can be of particular importance as poor conformation can dramatically affect her fertility. Look also for any scar tissue – perhaps from previous foalings in an older broodmare – the thickened fibrous tissue produced in response to injury can also affect the vulval seal. Check for any obvious signs of discharge, either around the vulva, or staining/scalding down her legs, as this will indicate a possible uterine infection.

Although it is another additional expense, it is vital to have any prospective mare examined by a veterinary surgeon for breeding soundness. This is a valuable examination whether the mare is going to be bred for the first time or if she has had many foals before. Maiden mares, going to stud for the first time, are an unknown quantity and the veterinary surgeon will examine the reproductive organs internally to ensure breeding soundness. Young mares may not yet be physically mature enough to be bred and the veterinary surgeon will be able to determine this during the examination. An older proven mare may have suffered more than just the normal wear and tear to her reproductive organs during her time at stud and this should be carefully evaluated before committing to a purchase.

Breeding horses can soon become very costly, especially if the broodmare requires several veterinary visits and treatments to assist her in conceiving. Stud fees for stallions can range in price from a few hundred pounds to many tens of thousands so reducing extra costs by selecting a mare in good reproductive health may not totally guarantee a good, healthy foal at the end, but does help to reduce the financial risks.

The Stallion

Choosing a stallion for a mare is very much subject to individual preferences and, in many cases, is also subject to who is 'fashionable' for that year. Stallions tend to go in and out of vogue depending on the recent success of particular bloodlines or progeny. A proven stallion is one who has already had the opportunity to prove his worth with his progeny. If they are successful, he will become even more popular and his stud fee may increase in line with his increasing commercial value. The value of first season stallions is usually based on their recent performance record, as well as that of their close relations, in addition to their pedigree. Many breeders avoid using a first season sire, as they can be an unknown quantity and therefore a risk, but for some this is the time they prefer as they may be able to use the stallion at his best price. Some of the universal factors that should be considered before coming to a final decision are:

- Purpose. For what purpose the mare is being bred; that is, to produce an offspring for future sale or for performance and so on.
- Financial. What money is available to spend on a nomination or stud fee.
- Pedigree. The pedigree of the mare should complement or enhance that of the chosen stallion.
- Conformation. If the mare has conformational faults then choose a stallion who does not have faults in the same areas.
- Location of the stallion. This factor is not so important if artificial conception methods are an option, as the mare need not be transported to the stallion's studfarm and thereby incur additional shipping and keep costs.

For a commercial breeder looking for a stallion to stand, the decision is normally governed by many of the above factors. It may be that the available stallion is homebred and, due to his success, is retired to stand on the breeder's studfarm, or perhaps he will be offered for stud duties whilst he continues to compete. Competing stallions are more common in the non-Thoroughbred breeds where artificial methods of conception are commonplace. For Thoroughbred stallions, artificial means of conception are, to date, banned under international ruling, which means that any Thoroughbred that is to be registered in the Thoroughbred Stud Book must be conceived by a natural mating. Additionally, due to the structure of racing and the extraordinary value of many Thoroughbred stallions, it is usually the case that they will retire to stud at the end of their racing careers.

There is no reason why a stallion cannot continue to work and compete as well as standing at stud, but there are practical considerations to take into account. If he must mate naturally with all his mares, it may be impossible for him to be away from the farm at all during the breeding season. There is also the very real risk of injury or infection: if he was unable to mate, either naturally or artificially, then considerable financial losses could be incurred, not to mention inconvenience to the owners who have sent mares to him. Temperament is another factor, some stallions do not cope well with doing both jobs at the same time, especially if competing at high levels and may require to concentrate just on their stud duties for the season.

The choice of the stallion is crucial for most commercial studfarms as normally it is the stallion, rather than the mare, that influences market prices. His success and popularity, pedigree and stud fee all influence the price that can be fetched for his offspring. Standing a stallion can mean that the owner breeder does not have to travel mares away and incur additional costs and risks as well as paying a stud fee for an outside stallion. However, owning and standing a stallion does have some disadvantages.

If the stallion is not homebred, then his purchase costs can be substantial. Added to this is the cost of promoting the stallion. It may be that the original investment is not repaid for some years and if the stallion is unsuccessful for whatever reason, then these costs may never be recovered.

Standing a stallion, almost regardless of breed, does require specialist facilities and experienced staff. There will be additional labour costs involved with teasing and examining the mares visiting the stallion. Taking in boarding mares from outside may result in extra income from keep fees, but will require additional stabling and/or grazing, as well as additional staff.

There are other options to purchasing a stallion outright, such as leasing or syndication. Leasing a stallion is usually done for a set period allowing a breeder to stand a horse that would not ordinarily be affordable. Syndication is common with Thoroughbred stallions and allows the breeder to buy an 'interest' in the stallion rather than buying him outright. Syndication usually involves several shareholders and, therefore, spreads not only the cost but also the risk. Shareholders pay for a share or shares, as well as an annual keep/administration charge. In return, shareholders are entitled to an annual breeding right for each share to use for their own mares, or to sell for someone else to use. Occasionally, a few of the shareholders in the stallion will not have their own mares and purchase the shares purely as an investment, selling their breeding rights each year for the current market value. The syndicate itself is generally managed by a bloodstock agency and can be quite a complicated affair, especially with very valuable stallions.

Before committing to setting up or expanding a breeding enterprise, it is important to look at the costs in detail. Converting or building the extra facilities that will be required for a stallion and boarding mares that come to visit him, may not be recovered immediately and the keep fees that it is possible to charge for the visiting mares may not cover the outlay on additional staff and feed, etc. Even in the case of a successful stallion, offspring may take years before they are saleable. The cost of producing even a foal for sale should be analysed realistically therefore.

Teasing

Teasing, or trying, is the term used to describe the management procedure for establishing whether or not a mare is in oestrus. The procedure is routine on studfarms where the stallion is kept separated from his mares during the breeding season. In the wild, the stallion is able to constantly assess the breeding state of his mares by smell and sight, getting closer to those that are receptive to him and further assessing their readiness for mating by taste and sound. The kept stallion, of course, does not have these opportunities so the teasing of mares is perhaps one of the most important aspects of stud management. Most mares display characteristic signs when they are in oestrus but this can vary to some degree, dependent on the time of year and the breeding experience of the mare. From a management point of view, one of the most important skills involved with teasing mares is observation. Experienced stud staff will constantly be looking at the mares, even when they are turned out in the paddock.

In simple terms, during the period that she is in oestrus or in season, the mare will be receptive to the stallion and permit his advances. When she is not in season, in dioestrus, she can be extremely aggressive to any approach. The normal teasing terms used on studfarms are 'showing', 'in use', 'on heat' to describe an in oestrus mare and 'teasing off', 'not in use' and 'not in season' to describe a mare that is in dioestrus. We have already learned in Chapter 1 about the physiological changes that govern these behavioural reactions.

Teaser Stallion

There are several ways of finding out which mares are in season, but the routine method is to use a 'teaser' stallion. The actual stallion himself may be required to do all the teasing on smaller farms, but due to the aggressive nature of mares that are not in season and the real risk of injury to the stallion, the role is more commonly carried out by an inactive stallion. On Thoroughbred studfarms, the use of a pony stallion for teasing is routine. The most important attribute of a teaser is that he is interested in the mares and that he has a good temperament. Good teasers are worth their weight in gold to a studfarm as they are often able to identify mares that are shy or difficult and do not display their status easily.

The use of vasectomized or sterile teasers used to be quite popular until the risk of cross-infection was fully understood. The main advantage of a sterile stallion was that he could run out with the mares and there would be no risk of them getting in foal by him; however, if one mare in the herd had a venereal (sexually transmitted) infection he would, quite naturally, spread it to all. Running a small stallion, such as a Shetland, with a group of larger sized mares is also a method that has been successful – the size difference naturally prohibiting mating. But this method has its disadvantages in that having a stallion, of any size, with mares presents difficulties in management and there have been cases where mating, however impossible it appears, has occurred. On most studfarms today, the teasers are kept in a similar way to the stallions themselves and quite separate from the mares.

Teasers may also be used to introduce a mare at stud for the first time to the mating procedure. In this procedure, known as 'bouncing', the teaser stallion is held on a bridle and permitted to mount the mare, but not to mate with her. Again, an experienced teaser is invaluable for this procedure as he will quietly mount the mare several times if necessary.

Although this sounds very frustrating for the teaser, experienced teasers effectively 'switch on and off' their sexual behaviour quickly, showing no interest in the mare as soon as they are turned away from her. Temperament of the teaser is vital, as already mentioned; not all stallions will tolerate the teasing procedure on a regular basis and some become disinterested with the whole affair and show no reaction to the mare, whatever her reproductive state.

Teasing Areas

The area that is used for teasing needs to be specially designed. Most studfarms have a designated area that is set up with safety for both horses and handlers in mind. Using a teasing board is the most standard method and this is formed by a reinforced and padded gate, or by a strengthened, padded area within a line of paddock fencing. It is imperative due to the violent reaction of mares that are not in season, that the structure is strong enough to stand the strongest of kicks. Conversely, it should also be high enough to deter an overly amorous teaser!

Teasing Methods

There are several different methods of teasing mares and it is common for variations to be used on any one farm. The standard method is for the teaser to be led to a specially designed teasing area and for the mares to be led to him, one at a time. This is the method that we will use when discussing teasing procedures later in this section.

Another method is to lead the mares, one at a time, past the teaser's stable where he can reach them over his stable door. If this method is used, the teaser's stable door should be adapted for this purpose and be reinforced, padded and increased slightly in height. Often on farms where staffing may be limited, a chute system is set up to allow the mares to go past the teaser without having to be led up individually. Teasing may also be practised by leading the teaser around the outside of paddocks where the mares are turned out. This is a useful method when young inexperienced mares are involved, as it allows them to be teased in a less stressful environment. The disadvantages are that it can be difficult to allow each mare the opportunity to approach the teaser, as the more dominant ones will tend to keep the quieter mares at bay. This method can also increase the risk of injury to the mares by aggressive mares kicking out. An adaptation of this theme is to tease the mares in the paddock by leading them individually to the teaser. This can be useful as it does away with the need to bring all the mares in to stables; however, there must be sufficient staff to hold all the mares away from the teasing area until it is their turn, again, to reduce the risk of injury to handlers and horses. To take this method one step further, some large studfarms, particularly in countries like Australia and the USA, have a teaser paddock within the mares' paddock. The teaser, normally of small pony breed, is turned out continuously in his paddock and the mares are free to come to him as they wish. This is only of practical use if the paddock sizes are big enough and the farm has more than one teaser. It is, again, essential that the teaser's paddock is designed specifically for the purpose with strong and high fencing.

Safety

Safety during teasing is paramount as the procedure can provoke very violent reactions from the mares. The teaser should be handled only by a person experienced with stallions. All staff should be aware of the teasing procedure, that is, where they should place the mare and where they should stand themselves. The mares should be led to the teaser on a bridle with a suitable bit and reins to ensure that the handler has complete control. If the mare is not under control, not only is it dangerous to the handlers but, also, the time spent with her at the board may be wasted by not being clear about her reactions. All handlers should wear suitable hard hats with the chinstraps fastened and gloves. Teasing should always be carried out in a safe area specifically designed for this purpose.

Teasing Procedure

During the breeding season, barren and maiden mares are normally teased every other day until their oestrus patterns can be established. (Barren is the term given to a non-pregnant mare who has been to stud before, but either failed to conceive last season or was rested. A maiden mare is a mare who is going to be mated for the first time.) Teasing after mating is discussed in more detail later in this chapter.

The mare is led up to one side of the teasing board and the teaser led up to the other. It is usual practice for the horses to be first presented head to head and then the mare is moved, if she is not aggressive, so that she is standing side on to the board. The teaser will usually sniff, lick and 'talk' to the mare, nipping gently at her neck and working his way towards her hindquarters. If the mare is in oestrus, she will be interested and not resentful of the teaser's approach, leaning in towards the board, straddling her hind legs and squatting slightly. She may progress in her display by producing a stream of thick, strong smelling urine, which is loaded with pheromones. Pheromones are Nature's chemical scent messengers, further signalling the mare's readiness for mating. Mares also use a specific visual signal to the stallion by lifting the tail, pulling back the vulval lips and exposing the clitoris, this signal is called 'winking'.

Further signs that the mare may be in oestrus may be noticed by her handlers during her routine care. Due to the behavioural changes associated with being in season, the mare may become quieter and easier to handle, this is particularly apparent in mares that are normally of a difficult temperament. Some mares may also display signs of oestrus when groomed over their hindquarters. Apart from observation one of the other most important and valuable skills for handlers involved in the care of broodmares is the ability to quickly get to know what is normal behaviour for each mare following their arrival at stud.

When she is in dioestrus, the mare will reject, often violently, every approach by the teaser or stallion. She may attempt to kick out or bite him, putting her ears back and clamping down her tail, leaving her handler in no doubt as to her status. However, some mares that perhaps lack experience, or are shy, may just become quiet when brought to the teasing board – showing no interest one way or the other. These mares need careful watching to avoid their behaviour being misinterpreted. Some of these mares may show signs of oestrus when put back in their stable and they feel more secure, or perhaps they will show to other mares out in the field. With modern commercial methods of stud management, such mares will rarely be missed as their status can be confirmed following an examination by the veterinary surgeon.

Mares that are pregnant, or in anoestrus, generally display dioestrus behaviour. Occasionally, mares may appear to be in oestrus although they are in fact pregnant. This can happen early in the pregnancy and the mare should be re-examined to ensure that she has not lost her pregnancy. It is very rare for a pregnant mare that is displaying oestrus to actually permit mating.

Abnormal Behaviour

Mares that are routinely teased during the breeding season come to associate the teasing area with the stallion and it is good practice to ensure that the mare is never forced to remain close to the teaser. There are cases of mares that always react aggressively towards the teaser, regardless of their sexual state and there are several reasons why this might occur.

It is important to remember that the teasing procedure on most studfarms is artificial. For many mares, being teased may be the only time that they have sight or sound of a male horse, apart from when they are mated; the noise and behaviour of a stallion can make mares nervous or defensive. For mares that have a foal at foot, the mere sight of a stallion can make them overwhelmingly protective of their foals and they may become very difficult to tease successfully.

It is not unusual for any horse to associate a previous bad experience with a certain place, person or procedure and to react badly when that place, person or procedure is encountered again – this is often termed conditioned behaviour. Teasing is an experience that can be stressful for some mares if it is not carried out properly and some mares develop a conditioned aggressive response to the teaser. Some of these mares can be so unpredictable that they can only be reliably assessed by veterinary examination.

Teasing Routine

Many studfarms prefer to tease their mares in the morning before the stabled mares are turned out. Teasing early in the day is particularly useful for farms which have regular veterinary attendance as the mares requiring examination can be kept in and the veterinary surgeon can be given a recent appraisal of the mares' teasing behaviour.

Once the mare has shown signs of oestrus, she will normally be examined and mated at the optimum time for her to conceive, that is, just before ovulation. The veterinary surgeon can establish the optimum time for conception to occur by examining the mare's ovaries when she is in oestrus. This is done by rectal palpation, the veterinary surgeon inserting a gloved hand into the mare's rectum and feeling the ovaries through the rectal wall. Experienced veterinary surgeons can tell the size of the developing follicles on each ovary, as well as predict when the mare is likely to ovulate. On large commercial studs, it is routine practice for the mare's ovaries and uterus to also be examined by ultrasound scanner prior to mating as this gives the veterinary surgeon a very accurate picture of the condition of the mare's reproductive tract. Once mated, the mare is then teased every other day until she shows signs of dioestrus, it is also common for the mare to be checked by the veterinary surgeon two days after she was mated to establish whether ovulation has occurred. The hormone progesterone secreted following ovulation means that the mare will start to display aggressive behaviour towards the teaser.

Once ovulation is established, the mare is normally left for 12 days from mating and then teasing her begins again, every other day. This may appear to be an early stage in her cycle to begin teasing again, but it is known for mares that may have a uterine infection to return to oestrus earlier than normal. Beginning teasing again at this early stage means that these mares can be identified and receive veterinary attention as necessary.

If the mare shows no signs of returning to oestrus after about 15–16 days from mating, she can be examined by the veterinary surgeon to establish whether she is pregnant. Veterinary use of an ultrasound scanner is now almost regarded as a routine method of diagnosing pregnancy in broodmares and this technology means that the pregnancy can be detected at a very early stage. Even though a mare has been diagnosed as in foal, it is normal for her to still be teased until she is about 28 days from mating, this is because some mares suffer early pregnancy loss and return to oestrus for no obvious reason. Combined with the continued teasing, it is common practice for the mare to be re-scanned at 21–28 days and again at 40–42 days from mating to ascertain that the pregnancy is developing correctly. (Diagnosis of pregnancy is discussed in more detail in Chapter 4.)

Mares who have foaled return to oestrus at about 6–14 days after foaling – this is called the 'foal heat'. It is thought that this period of oestrus is designed to allow the uterus to recover fully from the pregnancy, rather than for breeding; however, those breeds with restricted breeding seasons such as the Thoroughbred, are commonly bred on this heat. Mating on the foal heat will depend on the mare's recovery from foaling. The veterinary surgeon is normally able to assess her suitability for mating by checking the mare's uterus to establish if it has returned to its normal size and shape after foaling. A swab will also be taken from the mare's uterus to check for any signs of inflammation or infection. With modern veterinary techniques, mating on this heat does not normally affect the mare's chances of conception if her recovery from foaling has been good. Once the foaled mare has been covered, she will be teased in the same way as a maiden or barren mare.

The Problem Mare

The problem mare may be defined as a mare that has difficulty in conceiving and carrying a foal to term. The fertility of a mare is partly predetermined by the genes she has inherited from her parents, but also by the way that she is managed and the environment she lives in. The inability to conceive or the failure to carry a foal through to term are only the visible signs of a reproductive problem. We have already discussed the importance of understanding the 'mechanics' of the mare's breeding cycle and it is also helpful for any breeder to have a basic knowledge of the factors that cause reproductive failure. Problem mares that are left unidentified and untreated may progress from having simple reproductive problems to becoming permanently infertile.

In Chapter 1, we discussed the hormonal factors that control the mare's reproductive system, but this is just the start of the story. The uterus provides the growing foetus with protection and nourishment, allowing it to grow and develop until it is ready to be born. The uterus is, in many mares, susceptible to infection and inflammation – termed endometritis, literally inflammation of the endometrium, an impossible environment for the foetus to survive in. As well as the uterus, there are the main 'doorways' to the uterus, first, the vulva, which must seal off the genital tract from the outside environment, then the cervix, which must provide a complete seal to the uterus during dioestrus and, more importantly, pregnancy. If these 'doorways' do not shut properly, the mare will be prone to all manner of genital tract infections and will almost certainly be sub-fertile.

Susceptible mares are those that may be elderly, perhaps they have had many foals before and the structure of their genital tract is beginning to suffer from too much wear and tear. Sometimes young, apparently healthy mares suffer from repeated genital infections and fail to get in foal; these mares may have been born with, or acquired through injury, poor genital conformation.

However, the presence of bacteria in the mare's reproductive tract is not necessarily a disaster. Many of the organisms that can be found in the tract are environmental, which simply means that they are present in the environment all the time. These organisms do not affect the mare unless she is susceptible to them for some reason, such as if she is weakened by another infection, or if she has no immunity to them. To establish if the mare has any infection of her genital tract, it is necessary to take a sample of tissue and fluids, a process known as swabbing. In Thoroughbred mares, routine swabs are taken from the clitoral sinuses and fossas at the start of each breeding season and from the endometrium, via the cervix at each oestrus period. (Swabbing of mares is discussed later in this chapter.) Natural service by a stallion deposits a huge quantity of organisms into the uterus as the penis is quite a dirty object and penetration will also allow entry of dust-laden air into the genital tract. These organisms prompt a localized inflammatory response in the uterus, as the mare's immune system attempts to combat any infection.

Most healthy mares cope well and have fully recovered from the inflammation within a few days, so that when the fertilized conceptus arrives there is no remaining inflammation or infection of the uterine lining. However, susceptible mares, those that cannot deal with this infection so readily, are unable to prepare the uterus in time and the fertilized conceptus arrives into an inflamed, infected environment. The result being that the conceptus dies and the mare returns to oestrus, often earlier than normal due to the irritation of the uterus. This cycle will continue for as long as the mare is bred and her uterine infection goes untreated. Long-standing inflammation will ultimately damage the delicate lining of the uterus, the endometrium, which will result in scarring and this damage can be permanent. What makes some mares more susceptible than others is not completely understood, but several factors such as conformation of the external genitalia, age and previous infections are known to be contributory.

If the vulval seal is affected then air can be sucked into the genital tract, a condition known as pneumovagina. It does not just affect older broodmares, some types of mare are more prone to pneumovagina than are others: lean athletic maiden mares with minimal body fat may also suffer, as may mares in poor physical condition, or those simply born with poor vulval conformation.

Whether acquired or inherited, poor vulval conformation is usually characterized by a sunken vulva. The ideal conformation is that of an almost upright positioning of the vulva in relation to the pelvic brim and anus (Figure 3.1). The upright position helps to maintain the vulval seal and stop faeces from contaminating the vulval area. Mares with concave or sunken vulvas tend to catch faeces on the vulval shelf and combined with a poor seal, this permits faeces as well as air to enter the genital tract.

A common veterinary procedure to correct this problem is Caslick's operation, named after the surgeon who devised the technique. This simply involves stitching the lips of the vulva together to a point below, or in line with, the pelvic brim. A small area is left open so that the mare can urinate and the mare must have an episiotomy – 'be opened up', for mating and for foaling. Once a mare has been stitched in this way, she will require to be re-stitched after mating when ovulation is confirmed or foaling. Another, more complicated, procedure to correct severe cases is Pouret's operation, again named after the surgeon who first developed the technique. In simplified terms, Pouret's operation involves cutting horizontally into the muscular tissue separating the anus and vulva, literally drawing the tissue forward to correct the sloping position and effectively 'tighten' the seal. This type of technique is not used as often as a Caslick.

An associated conformational problem that can affect a mare's ability to conceive is when the vagina itself becomes sunken. A sunken vagina slopes down towards the cervix and, as urine is voided into the vagina just inside the vulva, gravity causes urine to pool in front of the cervix. A sunken vagina is associated with pneumovagina and is also common in mares with poor vulval conformation and body

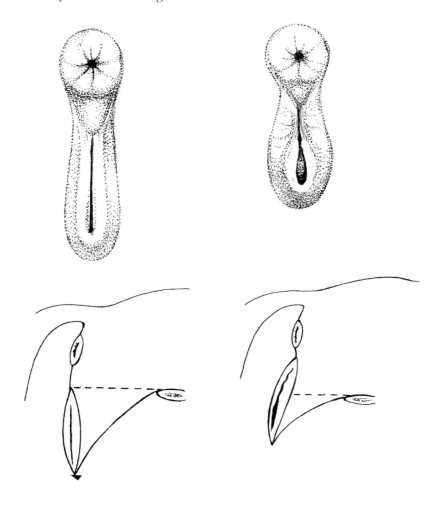

Figure 3.1 *Vulval conformation in the mare. Here, good vulval conformation is illustrated* (top, left), *an almost upright positioning of the vulva in relation to the pelvic brim and anus* (bottom, left). *Sometimes a mare will have a poor vulval conformation* (top, right) *showing a sloping vulval position of the vulva in relation to the pelvic brim and anus* (bottom, right)

condition. Mares that have recently foaled may suffer, temporarily, from a sunken vagina until their muscle tone is re-established.

If the mare suffers an injury or trauma to her reproductive tract, it may result in the formation of scar tissue and adhesions. This trauma may be caused by breeding or foaling injuries. Scar tissue of this nature may prevent proper closure of the cervix and predispose the mare to constant uterine infection. If the cervix is unable to close properly, it is unlikely that the mare will ever breed successfully.

In summary, the problem mare, one that is predisposed to genital infections, is unlikely to conceive successfully without costly veterinary assistance. Mares with mild endometritis will often be treated by the veterinary surgeon with antibiotics deposited directly into the uterus just after mating and this can be very successful. However, those mares with a long-standing problem that has gone undetected or untreated, may suffer permanent scarring to the uterine lining resulting in sub-fertility, difficulty in carrying a foal to term – and even if the foal is carried to term, it is likely that its growth and development will have been affected in some way.

Preventing Infection

Preventing infection on studfarms is one of the most important aspects of stud management, especially today when very large numbers of horses move to and from studs all over the world during the spring and summer months. We are all familiar with equine influenza and tetanus and most of our horses are vaccinated annually against these infections, however, in stud horses, other significant infections also need to be understood.

Most Thoroughbred studfarms work to a strict policy of routine vaccination for influenza and tetanus, EHV 1.4 and sometimes also blood testing and swabbing for all horses resident or visiting the farm. Each year in the UK, the booklet *Codes of Practice* is published, currently by the Horserace Betting Levy Board. The booklet outlines the recommended procedures for prevention of the main venereal diseases, as well as those such as strangles which, whilst not being a venereal disease, is extremely contagious and can cause severe problems to affected horses and farms. The *Codes of Practice* also details isolation procedures following a suspected outbreak of disease. In 2001, the outbreak of foot and mouth disease on agricultural farms in the UK illustrated the devastating effect of infection outbreak to the whole rural community, not just to those involved with farming. Although foot and mouth disease does not affect horses, and many of the diseases affecting horses are unlikely to have the same widespread effect, to the horse population and those involved with keeping horses, disease outbreak can cause massive losses. For this reason the *Codes of Practice* is published although, at present, not required to be followed by law and as recommended practice only for all studfarms. Sadly, there are many studs in Europe that do not insist on the simple tests for visiting or resident breeding horses that can help in disease prevention and, in consequence, there are cases of disease outbreak each year that could have been avoided completely.

The significant bacterial venereal diseases that affect breeding horses are Contagious Equine Metritis (CEM), Klebsiella pneumoniae and Pseudomonas aeruginosa. In addition, there are two viral conditions that are also of importance, Equine Viral Arteritis (EVA) and Equine Herpes Virus (EHV). A blood test is used to check for the presence of EVA and although EVA is not a venereal disease it can cause abortion in mares and is, therefore, significant. There is a vaccine available to protect against EVA, although at the time of writing (2002) it is not licensed for general use in the UK. Vaccination against EHV is readily available to breeders and it is common policy on Thoroughbred studs to insist that all pregnant mares are vaccinated. Pregnant mares are the most vulnerable group to infection by EHV as infection causes abortion without any obvious warning signs.

CEM, Klebsiella and Pseudomonas are tested for by swabbing certain areas on the genitalia of both stallion and mare. CEM is a notifiable disease and swabs must be cultured in a recognized laboratory. All of these infections cause inflammation of the mare's uterus, which can be prolonged and, because the endometrium becomes irritated, the mare may come back into oestrus earlier than normal. If mated, the mare will transmit the infection to the stallion who will then infect any other mare that he is mated with. CEM is very contagious, it is also possible for it to be transferred from horse to horse by stud staff, so stable hygiene is very important. Once infected, some mares may become carriers showing no signs of infection themselves. Stallions can also harbour infection with no outward sign. Although CEM does not cause abortion, any foal born to an infected mare will also harbour the bacteria in its genitalia. The clitoral fossa of the mare and the prepuce and urethral fossa of the stallion all contain smegma; in this warm, dark, moist environment, bacterial infections flourish and these areas are specifically tested for venereal pathogens.

The main method of preventing and identifying venereal disease on studs is to annually swab all broodmares and stallions. In the mare, this is done by taking two swabs, one from the clitoral sinuses and fossa and one taking cells from the lining of the uterus, an endometrial swab. Clitoral swabs are taken at the start of the breeding season and must be sent to a recommended laboratory for culture. A repeat clitoral swab should be taken if the mare is to be mated by another stallion during the same season. Endometrial swabs are taken when the mare is in oestrus as it is necessary for the cervix to be open to allow access to the endometrium; these swabs are then repeated on every heat period that the mare is mated. Reputable studs will require proper veterinary certification for each swab to confirm that no disease is present. The stallion is swabbed at the beginning and sometimes also at the end of the breeding season. Swabs are taken from the prepuce, urethra and urethral fossa. Again, these swabs must be sent to a recognized laboratory for culture.

Most Thoroughbred studs require an additional form to be completed by the mare owner which details the mare's breeding history for the previous three years (Figure 3.2). This information is important to the stallion studfarm, especially if the mare in question has been on a studfarm known to have had an infection outbreak, or if the mare herself has had a positive swab for these bacteria within this timespan.

EVA is a viral condition that is a significant cause of abortion. Spread during mating, EVA can also be transmitted through insemination with infected semen and may also be spread by droplet

CONTAGIOUS EQUINE METRITIS
AND OTHER EQUINE
BACTERIAL VENEREAL DISEASES
(2002 SEASON)

MARE CERTIFICATE

Certificate to be completed by mare owner and lodged with the prospective stallion stud farm owner before the mare is sent to the stallion stud farm.

Name of mare ....

Passport number (where available)..

Name and address of owner...

....

....

1999 stud visited..

mated with ....result ....

2000 stud visited..

mated with ....result ....

2001 stud visited..

mated with ....result ....

Additional information including the results of positive bacteriological examinations for CEMO, *Klebsiella pneumoniae* and *Pseudomonas aeruginosa* at any time:

....

....

NAME (PLEASE PRINT) ....

SIGNATURE ....

DATE...

Figure 3.2 *A CEM form detailing a mare's breeding history for three years. (Reproduced by kind permission of the Horserace Betting Levy Board.)*

infection (coughing and sneezing, etc.). Whilst mares recover very quickly from EVA infection, some stallions become carriers and permanently 'shed' the virus in their semen. The prognosis of a 'shedder' stallion is normally very grave. All horses can be blood tested for the presence of EVA. In simple terms, the blood sample is tested for the presence of antibodies to the virus; if the horse has been in contact, or is currently infected, with the virus there will be antibodies present in the bloodstream.

However, a horse with a positive blood sample may no longer have the virus so repeat tests are taken to see if the level of antibodies remains the same. A rising level of antibodies signifies that the horse currently has the virus and is infectious, a static level of antibodies means that the horse has been in contact with the virus at some time, either by vaccination or infection, but is no longer infectious. The *Codes of Practice* recommends a blood test for EVA to be taken every year from every horse visiting or resident on a stud. For those horses that have travelled to stud in the UK from countries outside Europe, additional tests are required, as with those who have shown a positive blood test. The recommended frequency of these tests is detailed in the *Codes of Practice*, but is generally in the region of one test followed by a further test 14 days later. It is very important that these horses remain in isolation until the required tests have been completed.

EHV is a relatively common virus amongst the horse population in the UK and, in its respiratory form, commonly affects foals and yearlings. There are four types of herpes virus, EHV1 being the most significant to broodmares as it is the single most important cause of abortion. (EHV1 is discussed in detail in Chapter 5.) EHV3, also called Coital Exanthema or 'the Pox' (Figure 3.3) does not cause abortion in mares, but is spread by mating and by poor stable hygiene. Affected mares have blister-like lesions on the vulva and surrounding genital area. Stallions develop similar blister-like lesions on the penis. The blisters rupture and produce ulcers that may become infected. The condition will clear up of its own accord, but may require antibiotic treatment to deal with secondary bacterial infection of the ulcers. EHV3 infection tends to occur erratically and it is often difficult to be certain with which horse it began as it may also be spread by handlers and contaminated veterinary equipment. Vaccination against EHV 1.4

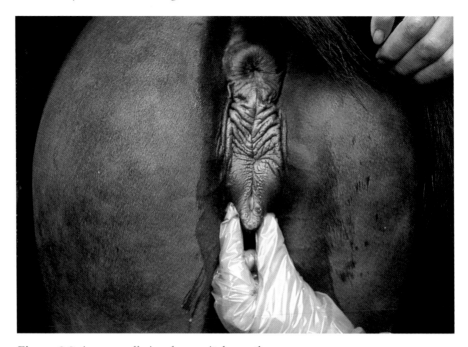

Figure 3.3 *A mare suffering from coital exanthema*

is available to all horse owners and it is the policy of Thoroughbred studs to insist that pregnant mares are vaccinated if they are to foal on the stallion stud. There are now also combined vaccines available to protect against both equine influenza and herpes virus, although these vaccines are not currently licensed for viral abortion.

Strangles has recently been included into the *Codes of Practice* as there have been several outbreaks on Thoroughbred studs. Strangles is not a venereal disease, but it can cause major problems to any breeding operation because it is so contagious and an outbreak will mean that the whole stud will be put into isolation. Strangles is a bacterial infection that affects the respiratory tract and lymph nodes of the horse, causing nasal discharge and abscesses that form around the jaw and throat. The abscesses will burst after a few days and produce thick pus. Affected horses will have a temperature and shows signs of discomfort when eating, although this is commonly relieved once the abscesses have burst. A complication of the infection is often called 'bastard strangles', this is when the abscesses develop and burst internally. Bastard strangles may cause serious complications. Strangles is diagnosed by taking swabs from the nasal passages of the horse and it is possible for horses to become carriers and shedders of the bacteria without developing any clinical signs themselves. Identifying a carrier is of the utmost importance and this is why all horses on an affected farm should be swabbed, not just those showing signs of infection. (Isolation procedures following an infection outbreak are discussed further in Chapter 5.)

Breeding Methods

There are four main methods used for breeding horses:

- Paddock or free breeding, when the stallion runs free with his group of mares in a large paddock or, perhaps, as a semi-wild herd such as on the New Forest, Dartmoor or Exmoor.
- Artificial insemination, where semen is collected from the stallion and the mare is inseminated rather than mated naturally.
- In-hand breeding, which is currently the most common method on commercial studs, where the stallion and mare are brought together only for mating and each is held by handlers throughout the mating process.
- Embryo transfer, when an embryo is taken from one mare and implanted into another mare who will carry the pregnancy to term.

The more modern of these four methods, and becoming increasingly prominent, are artificial insemination and embryo transfer. Artificial insemination (AI) is, as its name suggests, one of the most artificial methods of breeding. It involves collecting semen from the stallion using an artificial vagina and then inseminating a mare with the prepared sample. Embryo transfer (ET) is also an artificial method of

breeding and it may be performed by insemination or surgical procedures. For ET, the mare is either bred naturally, or inseminated and
the resulting conceptus is literally flushed out of her uterus about
eight days after mating. The collected conceptus is then implanted
into another mare, called a recipient, who will carry the pregnancy to
term – the equine form of surrogacy. AI and ET are currently not permitted by all breed societies, notably so with the Thoroughbred.

Correctly managed, both AI and ET dramatically reduce the risk of
disease and injury to breeding horses. The mare does not need to
travel to the stallion stud and the mare owner, therefore, is not only
able to keep her at home but also does not incur the expense of additional transport, keep and veterinary fees whilst she is away at stud.
Stallions that are used for AI and ET can often remain in competition
throughout the year, collections being timed to fit in with their training commitments. With ET, high-level performance mares also can
remain in competition whilst another mare, or mares, carries her foal
or foals to term; this arrangement means that the foals are born at a
time when the mare is at her highest market value. Additionally, those
mares that have suffered an injury that makes natural mating and/or
pregnancy impossible are still able to be bred from either by AI or ET.

We look at both AI and ET and more fully discuss in-hand breeding
later in this section, but begin with what might be called the traditional method of breeding horses.

Paddock or Free Breeding

Allowing the stallion to run with his mares is the most natural method
of breeding. The stallion and mares are able to behave as they would
do in the wild state, as long as the area they are turned out in is large
enough for the size of the herd. It is not a method that is widely practised by commercial studs as the management disadvantages tend to
outweigh the advantages.

The main disadvantages are that it is not possible for the mares to
be managed individually in this situation and, therefore, almost
impossible to be certain which mares have been mated and on what
dates and which mares are in foal.

The risk of injury to valuable breeding stock is very high, as is the risk of infection even if all the horses are thoroughly checked prior to their introduction to the herd. It may be difficult to introduce new mares to an established herd and many mare owners may not want their mares to be bred in this way because of the risks. Checking the horses can only be done by observation as some stallions object to the presence of people within the herd and any sick or injured horses are, therefore, difficult to treat.

However, a well-managed paddock breeding stud can be ideal for some small breeders and produce good levels of fertility. If there is sufficient space for the herd to roam and there is a small number of permanent mares with the stallion, this form of breeding does greatly reduce the management stresses on the horses. The close proximity of the stallion to his mares at all times tends to encourage the mares to cycle regularly and also encourages those mares that have been shy or unreceptive to become more relaxed. The other main advantage is the reduction in keep and labour costs to the stud owner as well as reducing the need for specialized stabling, breeding and veterinary examination areas.

Artificial Insemination

A very successful method of breeding, AI has been used extensively for many years with many of the agricultural breeds. However, it is also very technical and requires a high degree of expertise and veterinary assistance to produce good levels of fertility. Most stallions can be trained to use an artificial vagina (AV), which is essentially a water-filled cylinder with a collection bottle at the end. Training a stallion to use an AV can take some time during the initial period, but, with care and expert handling, the horse may become very relaxed with the procedure. Collecting the semen from the stallion is usually carried out by allowing the stallion to mount a mare that is in oestrus and then deflecting the penis into the AV rather than allowing him to penetrate the mare. Alternatively, a phantom or dummy mare can be used (Figure 3.4). Phantom mares are common on studs that specialize in AI breeding.

A phantom mare is essentially a padded stand. The stallion is taught to mount it and then ejaculate into the AV. It is relatively easy to train most stallions to use a phantom mare and after a while many become aroused just being led near to it, just as they would do when being led to the breeding shed for natural mating. Some stallions, however, will not accept a phantom mare and will only allow collection if they are able to mount a mare that is in oestrus.

Figure 3.4
A phantom or dummy mare being used for an artificial insemination (AI) collection

Studs that regularly collect from their stallions for insemination will also have laboratory facilities to enable the collected ejaculate to be checked and prepared correctly. Sperm is very susceptible to temperature changes and will quickly die if allowed to get too cold or too hot without the proper preparation. The preparation techniques for collected ejaculate will vary depending on whether the ejaculate is to be inseminated into the mare immediately, to be chilled and transported, or to be frozen for storage.

Correct preparation, storage and insemination of semen is essential and requires skilled staff and specialist facilities, the specific technical details of which are beyond the scope of this book. However, the principle is that the complete ejaculate is separated, usually using a centrifuge. This is to ensure that the sperm portion is removed from the seminal fluids and gel which may actually harm the sperm after ejaculation (see Chapter 2). The separated sperm portion is kept at the correct temperature throughout the early preparation and will be examined to make sure that it is of sufficient quality. Semen extender is then added to the sample; this is, in basic terms, nourishment and protection for the sperm. Depending on the number of sperm present in the prepared sample, it is often possible for it to be split into several portions. In theory, each of these portions can be used to inseminate a mare successfully, but, in practice, more than one portion is commonly used for each mare. Each of the portions is put into a special insemination tube called a straw. If the mare is to be inseminated immediately, the ejaculate can be either transferred to her whole, or separated. If necessary, it is also possible for antibiotics to be added to the sample for insemination and this procedure may be used with mares that are susceptible to uterine infection following mating. The mare is actually inseminated by the veterinary surgeon placing the contents of the prepared straw through the cervix into the uterus using a syringe. This must be done with great care as the sperm are vulnerable to damage as they are pushed out of the straw. Veterinary surgeons carry out insemination under minimum contamination procedure; this, to prevent the introduction of any infection, such as from dust, etc., into the uterus at the same time as the sperm is deposited.

If the sample is not to be used immediately, the straws can be chilled or deep frozen. Chilling and freezing are both complicated procedures that require specialist facilities and knowledge to ensure that the sperm are not damaged in the process and remain able to fertilize the mare when revived. Chilling sperm for insemination has very good fertility rates but, to date, freezing equine sperm has not proved as successful. However, the advancements in modern medical and veterinary science are constant and this situation is changing daily. In a few years, it may be commonplace for frozen sperm from long-dead stallions to be used as a matter of routine and being able to chill or freeze sperm means that the sample for insemination can be shipped to a waiting mare anywhere in the world.

In-hand Breeding

By far the most common breeding method practised on commercial studs, in-hand breeding involves the mare and stallion being brought together just for the act of mating and both will be held by handlers throughout. This method is particularly common on studs that breed Thoroughbreds as any form of artificial conception will prohibit the resultant offspring from being registered as a Thoroughbred. Thoroughbred stallions also tend to be amongst the most valuable of all stud horses and in-hand breeding means their management and that of their mares can be fine-tuned, minimising risks to health and fertility.

In-hand breeding is normally carried out in an area designed just for this purpose – on many Thoroughbred studs, breeding barns can be quite palatial. However, the basic principle for the breeding area is that it should be safe for both horses and handlers. (The layout of breeding areas is discussed in more detail in Chapter 8).

Stallions on commercial studs can have in excess of 100 mares booked to them, all to be mated during, in a number of cases, a restricted breeding season. There is wide individual variation in the number of mares that a stallion can successfully mate with and this number is greatly dependent on individual fertility, as well as the quality of the management of both stallion and mare. Some stallions are able to mate three, possibly four times a day every day with no adverse effect, others barely manage once a day on a regular basis without a drop in their fertility. In the past, when routine veterinary assistance was minimal and many of the modern diagnostic techniques were not available, it was likely that a stallion would not have more than 40 mares booked to him for the season. Today, high-level Thoroughbred stallions may mate with as many as 200 mares – some popular stallions even travelling to countries such as Australia or New Zealand to be used for the southern hemisphere season as well – essentially these stallions are mating all year round.

The careful management that the mare receives prior to her mating is so that she will have the best chance of getting in foal from just one mating. She is examined by the veterinary surgeon regularly during oestrus and only brought to the stallion when she is at the optimum point to conceive. Once mated, she will be examined again, usually 48 hours afterwards, to ensure that she has ovulated. With so many mares needing to be mated by just one stallion, it is important that matings are not 'wasted'. If mares need to be mated three or four times before conceiving this can effectively treble a stallion's book of mares and may result in his fertility being affected through over-use.

Embryo Transfer

Embryo transfer (ET) is a relatively recent advance in equine artificial reproduction methods. This technique is still not permitted for use with every breed society, but is becoming almost routine with some performance horse breeds. The beauty of embryo transfer as compared to ordinary insemination, is that the mare whose egg is fertilized does not carry her foal to term as this role is performed by a surrogate, or recipient mare. This has obvious advantages as we saw earlier, the mare is at the peak of her career and can continue to compete so that the market value of her offspring is potentially at its highest. The use of ET also means that it is possible for the mare to have more than one foal each year: in theory, there is no real limit to how many embryos can be fertilized and then transferred from the donor mare to any number of recipients. Another major advantage is that the use of recipient mares means that a mare that may be physically unable to carry a pregnancy can still have a foal. As with all

being experienced enough to carry out each step calmly. Carefully planned emergency procedures should be drawn up by the stud or stallion manager or the person responsible for the stallions and each breeding area attendant should know what action to take if a problem should occur. Many commercial studs will not permit staff other than those involved with the mating into the breeding barn, but some have viewing areas provided for visitors and mare owners.

Preparation

The stallion may become more highly strung and boisterous when in the breeding barn and should be held on a bridle with a suitable bit and long lead rein. Some stallions require a stronger bit than normal for mating to ensure that the handler has complete control throughout. Mares should also be held on a bridle and it is usual practice for the mare to have two attendants, one at either side of her head. All staff in the breeding area should wear correctly fitted hard hats.

In the past, it used to be common for the genitalia of the stallion and mare to be washed with soap or disinfectant prior to the actual mating, but this is now known to be detrimental. Soaps and disinfectants will kill the natural protective bacteria on the genital organs and allow infection by harmful bacteria. The use of harsh detergents and washing procedures can also cause the sensitive skin to become inflamed and, again, allow secondary infection to occur. The external genital organs of the mare and stallion are now normally washed, if at all, with just plain warm water to remove the majority of any faeces, dust and smegma.

The mare will have a tail bandage put on to help keep her tail hairs away from the stallion's penis. Studfarms that have a large number of visiting mares each season will tend to use disposable bandages that are thrown away after one use, but a normal tail bandage is adequate if it is clean and washed well after every use.

Once prepared, the stallion and mare are brought to the breeding area. It is usual practice for the stallion to tease the mare for a short period prior to the actual mating. The length of time required for teasing depends on the individual stallion and the status of the mare. A young maiden mare may require longer teasing prior to being mated so that she has time to relax – stallions can become very vocal and excited when approaching a mare for mating and some inexperienced or nervous mares may become afraid. Teasing also encourages the stallion to 'draw' (extend his penis and produce an erection) and become fully aroused. The length of time required for the stallion to produce an erection also varies enormously, some stallions achieve an erection just by being led towards the breeding area, others require quite a long period of teasing to become sufficiently aroused. It is common for young stallions to require a longer period of stimulation, particularly during their first season. However, a stallion that is slow or reluctant to arousal can be frustrating not only to the stud staff but also to the mare owner. Stallions with this type of problem, whether it is physical or psychological in origin, are very time-consuming and often produce poor fertility rates each season.

Preparing the Mare

Once the mare has been teased, she will be led to the breeding area to stand for the stallion. It is usual for busy studs to have a set place for the actual mating. This is often designed with a mound or dip built in so that the size difference between mare and stallion can be accommodated (Figure 3.5). A large mare being bred to a smaller stallion can be put to stand in a dip and a smaller mare being mated by a larger stallion can be stood on a mound.

Once positioned, the mare will have hind covering boots put on (Figure 3.6). These are thick felt or leather boots specifically designed to protect the stallion if the mare should kick. A mare in full oestrus is unlikely to resent a stallion, but nervous or defensive mares can sometimes kick out at a stallion when he tries to mount or dismount. Mare owners are always required to have their mare's hind shoes removed prior to mating, but the full force of a kick, even with an unshod hoof, can cause considerable damage (Figure 3.7).

Injury from a kick during mating can result in the stallion having to be rested from stud duties for some considerable time and, in the worst cases, can cause permanent damage. The boots completely

Figure 3.5
A platform is often built in to the breeding area to accommodate a size difference between mare and stallion

Figure 3.6
A mare prepared for mating, with hind covering boots, neck cape and tail bandage

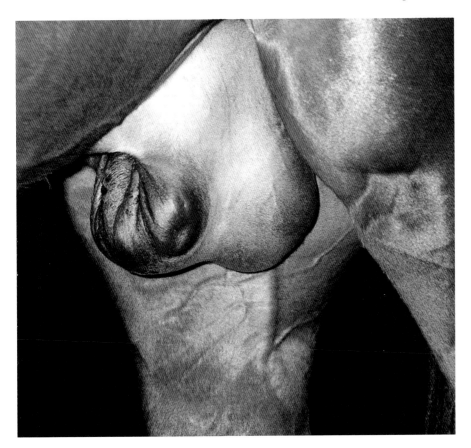

Figure 3.7 *Swelling of a stallion's sheath and testicles resulting from a kick during mating*

cover the hind hooves of the mare and should fit securely so that they cannot slip or come off. Maiden mares may become upset at the sensation of the boots and may need to be walked around a few times to become accustomed to them.

Some studs may also use a cape or neck rope. The cape is made of thick leather or canvas and completely covers the mare's neck, fastening underneath. Neck capes are purely protective and are used on mares being mated by stallions that may bite or savage their mares during mating. The neck rope is used for similar reasons and is a thick padded leather rope that the stallion can bite during mating. The use of equipment like this will vary depending on the character of the individual stallion and may not be necessary at all. On busy commercial studs, each stallion may have his own set of equipment for the breeding barn; this is not an extravagance as stallions are particularly sensitive to scent and may reject a mare who is wearing equipment smelling of another stallion.

The use of a twitch placed on the mare's nose is still very common during mating. (Although care should be taken when using a twitch with a recently-foaled mare. See Chapter 6.) The twitch is applied to ensure that the mare remains calm and still for the stallion and is removed as soon after mating as it is safe to do so. Twitches should only be applied at the last minute, especially if the stallion is slow to cover. The mare's attendant will hold the lead rein and the twitch handle, but if two attendants are used, one either side of the mare's head, then one will hold the reins and the other the twitch.

Hind leg hobbles were once commonly used during mating. However, if the mare is correctly prepared and in full oestrus, the use of hobbles is unnecessary. Hobbles can be dangerous if used by inexperienced handlers and can cause the mare to fall. However, a strap used to hold up one of the mare's forelegs is still relatively common. This is put on around the mare's pastern and is used to ensure that she cannot move forward as the stallion mounts. It is used most often with maiden mares who may jump away from a stallion as he tries to penetrate her as well as with young inexperienced stallions who may be unbalanced by the mare moving whilst they are trying to mount. The strap must be released so that the mare can stand on all four legs as soon as the stallion has penetrated her. The mare should always be allowed to take a step or two forwards during mating especially if the stallion has a large penis. Allowing her to move forward reduces the risk of her cervix becoming bruised with the stallion's thrusts.

Breeding rolls (Figure 3.8) may also be used when a large stallion is mating with a small mare, as may be common with, for example, a warmblood to a Connemara. A breeding roll is a large, normally leather, padded cylinder with a handle at one end. It is held by an attendant against the buttocks of the mare and the stallion's thighs to

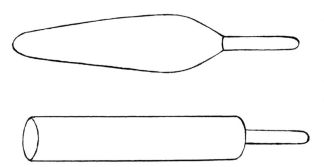

Figure 3.8
Two types of breeding roll used to reduce the risk of the mare's cervix being bruised during mating

stop him from thrusting too deeply. The cervix has a delicate structure and any damage will reduce the mare's chance of conceiving, possibly reducing her chances permanently.

Coitus

Once the stallion is stimulated, it is important that he is allowed to mate with the mare as soon as possible to avoid him becoming too excited, or losing interest. The stallion handler will lead the stallion up to the mare's left side at an angle of about 45 degrees. Inexperienced stallions may try to mount immediately when brought to the mare, jumping towards her from the rear. Mounting from one side reduces the risk of the stallion being injured by the mare if she should kick out. Mature stallions will approach their mare quite calmly and this quiet approach should be encouraged in young stallions. Approaching from one side also allows the mare to see the stallion coming and, in full oestrus, she will often further signal her receptiveness by squatting and urinating, as well as leaning towards him as he mounts.

The mare's handler should stand to the side of her head and just in front of her shoulders. Standing just to one side of the mare means that the handler can move out of the way should the mare strike out with a foreleg. This handler is responsible for restraining the mare and for moving her quickly out of the way if necessary.

The stallion's handler should stand to the left side of the stallion and maintain control at all times. Some stallions strike out with their forelegs as they mount and can inadvertently catch an inattentive

Figure 3.9
Breeding in-hand. Each attendant is wearing a properly secured hard hat for safety and the stallion's attendant is ensuring that the lead rein does not become caught on the stallion's legs during the act of mating

mare handler with a hoof, or get a foreleg caught in their lead reins as they move forward onto the mare (Figure 3.9).

There is normally another attendant, usually the stud or stallion manager who stands to the right side of the mare at her tail. This attendant is responsible for holding the mare's tail out the way as the stallion mounts and for guiding the stallion's penis into the mare's vagina if necessary. This attendant is also responsible for checking that the stallion has ejaculated properly during the mating. This is confirmed by the attendant placing a hand on the underside of the stallion's penis and feeling for the pulse of seminal fluid along the urethra as the stallion ejaculates. Stallions do display more obvious signs of ejaculation, such as deeper thrusts and tail 'flagging', but feeling for the pulse is the only way to be absolutely certain that ejaculation has taken place.

It may sound a little strange, but some stallions become adept at 'faking' and may only pretend to ejaculate. There may be a number of reasons why the stallion exhibits this behaviour but it is vital that a failure to ejaculate is identified as early as possible. One further point to mention is that the excessive handling of the stallion's penis or legs during this stage of mating should be avoided as almost all stallions resent too much interference.

The number of thrusts required before the stallion reaches ejaculation varies between stallions and time of year. The actual mating lasts, on average, about one minute, but may be as short as 30 seconds, or as long as five minutes.

Once ejaculation has occurred the stallion should be allowed to stay on the mare for a few moments as, normally, he will dismount of his own accord. The mare should be turned to her left to avoid her being able to kick out at the stallion as he dismounts; the stallion is then backed and turned away from the mare. Allowing the stallion to dismount too quickly immediately after ejaculation should be avoided as the glans, or head, of the penis becomes enlarged during ejaculation to assist in transferring the sperm directly into the mare's uterus. If the stallion dismounts before this has subsided, it can cause bruising and cause air to be sucked into the vagina.

After mating, the stallion's penis is normally washed with plain warm water. His front legs and belly may also be washed to remove the smell of the mare and any sweat. The mare will have the covering boots and tail bandage removed, as well as the neck cape or rope if this has been used.

Some breeders like to walk the mare for 10–15 minutes after mating to stop her straining to urinate and expelling semen. This is not necessary as the semen will have been deposited directly into the uterus in a proper mating and little, if any, semen should be present in the vagina. Any semen that could be expelled at this stage is of no use to the mare.

If the mare to be mated has a foal at foot, the question of whether or not her foal should remain with her in the breeding area is very much subject to the policies of the stallion stud. Some studs will not permit the foal to remain with the mare under any circumstances and it will be left with an attendant in the stable or horsebox. Other studs permit the foal if the mare is likely to be calmer. Some mares become distraught if they are taken away from their foals and will not permit the stallion to approach whereas others show no concern whatsoever. If the mare is very protective of her foal, 'foal proud', then allowing the foal to stay with her is likely to keep her calmer for the mating.

Administering tranquillizers for mating is normally only carried out in extreme circumstances. Mares that are unwilling to stand for a stallion should always be re-examined by the veterinary surgeon and the use of sedation should be considered only as a last resort. As was discussed in Chapter 1, a normal mare at the correct stage of oestrus for conception will be very receptive towards a stallion. Tranquillizers should only be given by a veterinary surgeon who should remain present throughout the mating in case of an adverse reaction. The big disadvantage with administering such drugs is that the mare is usually very upset by the time the drug has been given and this will affect the reaction.

4 | Pregnancy

Throughout the mare's pregnancy, extraordinary changes occur to the newly fertilized egg. Gestation, the period from conception to birth, takes about eleven months (340 days). The fertilized egg is just visible to the naked eye and is usually about 0.1 mm in diameter; astounding to think that an animal as large as a horse will develop from such a tiny object.

The egg is small but the foetus developing inside the mare (Figure 4.1) does not need to rely on its own food supply, as it would if it were a bird – which develops by feeding from a large yolk sac inside the egg. The horse foetus develops by receiving its nourishment from the mother's uterus throughout gestation and, although it does have a small yolk sac to feed it through the first month, this is soon replaced by the complex structure of membranes and blood vessels, called the placenta, which transfers nourishment from the mother to the foetus.

Before fertilization the egg contained 32 chromosomes; sperm cells also carry 32 chromosomes and when fertilization has occurred the conceptus will contain 64 chromosomes, half from the mother and half from the father. This number is only found in sperm and egg cells; in other cells of the body, 32 pairs of chromosomes are found. The combining of the chromosomes from both mother and father starts the intricate process of development from egg to fully formed foal. Each of these chromosomes is essentially a long string of proteins; each string has a pair and on each of these strings there is all the information, in code form, that will tell each cell what to do, where to be and

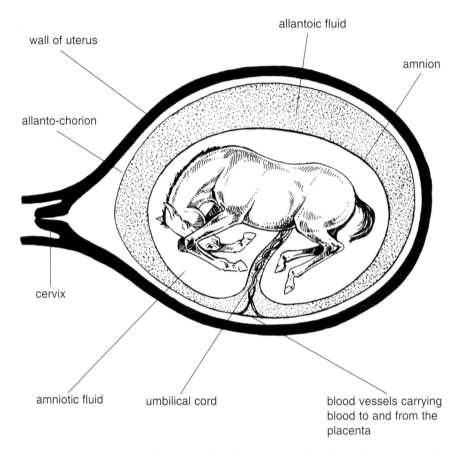

allantoic fluid

wall of uterus

amnion

allanto-chorion

cervix

amniotic fluid   umbilical cord   blood vessels carrying
blood to and from the
placenta

Figure 4.1 *Diagram to illustrate the placenta and foetal membranes that form during pregnancy to support the developing foetus*

when to divide as well as influence the characteristics of the foal. The information on each of the chromosome strings is made up of a set number of packages called genes. When choosing which stallion to mate with which mare, breeders are looking to reproduce the parents' genetic potential with the resulting offspring.

The genes that each cell possesses may also be termed a dominant or recessive. If we accept that each chromosome has a pair and that on each chromosome are located the genes, it follows that each gene should also have a pair that fulfils a similar, if not the same, function, for example, the gene for coat colour.

How the horse looks, its outward appearance, is termed the phenotype; internally, the type of genes the horse possesses in each cell is termed the genotype. From looking at a horse, a breeder can always tell the phenotype, but not the genotype as the horse may have inherited any number of recessive genes. Aside from the parents, there may also be inherited traits passed down from the grandparents and great grandparents; it is not unknown for some traits to skip one or two generations completely and then reappear. This is another reason why great care should be taken in researching the pedigrees of both mare and stallion before a final mating choice is made.

The current work on the human genome means that in the not too distant future it may be possible for humans to breed horses on the basis of their genotype, rather than just on their phenotype. But, although the ongoing genetic research indicates that anything may be possible, at present the genetic combinations that are achievable when mating one horse to the next are so enormous that, however much care is taken with the choice of sire and dam, there is always an element of chance in how the foal will turn out.

Research has shown that there is a gene for twining, essentially this means that a mare carrying this gene will produce twin ovulations more often than a mare who does not have the gene. With today's high level of veterinary supervision and available technology, twin pregnancies are removed at a very early stage by veterinary intervention (as described later in this chapter).

Whereas, in the past, a mare that produced twin pregnancies was often deliberately removed from the breeding herd, as mating would commonly result in aborted twins, or in the mare producing dead foals close to term. The disadvantage today is that mares with the twinning gene remain in the breeding herd and any filly foals born to them will also carry the twinning gene.

Research carried out in Thoroughbreds by the Equine Fertility Unit in Newmarket has shown an increase in the number of twin conceptions produced from 3.8 per cent in 1982 to 7.8 per cent in 1992.

The most common form of twin pregnancy is that where the mare ovulates on two follicles on an oestrus period and both are fertilized. This will result in non-identical twins. Less than 10 per cent of these pregnancies will produce live foals at term.

In 2002, the first proven case of genetically identical twins in the Thoroughbred was found (Figures 4.2 and 4.3). The Thoroughbred mare Calabria, belonging to Queen Elizabeth The Queen Mother, foaled two colt foals in February 2002 at the Royal Studs, Sandringham in Norfolk. The colts were the product of a mating to the stallion, Zaffaran in 2001.

The foals were born normally at full term. The fact that the mare had carried twins came as a complete surprise to all concerned as the mare had no previous history of twin conceptions and the presence of twins had not been recorded during the mare's routine examinations for pregnancy. The foals were healthy at birth, which is unusual for twins, although one was smaller than the other. It was not until the foals were DNA tested for registration in the Thoroughbred Stud Book that it became apparent that they were genetically identical. As this was the first recorded case in Thoroughbreds, further detailed examination of the foals' DNA was carried out by the Animal Health Trust to confirm the initial findings. Unlike identical twins in humans, the horse twins do not look alike – the white markings and coat whorls are completely different on each individual. However, this is not entirely unexpected, as it is already known that such markings are not strictly dictated by the parentage of an individual unlike coat colour.

(Professor 'Twink' Allen of the Equine Fertility Unit, Newmarket has published work involving a case of identical triplets in the horse as well as recording a case of identical twins in a Suffolk mare.)

Relatively recent equine research has enabled us to understand so much more about what is going on inside the pregnant mare's uterus for those eleven months of gestation. Detailed study of the developing embryo is beyond the scope of this book and beyond the needs of most breeders; however, an understanding of equine embryology in general terms can assist in planning the best management for the pregnant mare as well as for her developing offspring.

In the following sections of this chapter on pregnancy, therefore, we will look briefly at those areas most likely to warrant the attention of

Figure 4.2 *Identical twin colt foals born at The Royal Studs, Sandringham, in February 2002, shown here with their dam*

Figure 4.3 *The twins are identical, but differ in appearance, in fact even their coat whorls are different. However, both foals share exactly the same genetic information as they were naturally produced by the splitting of a single fertilised egg*

those people who are engaged, in whatever capacity, in the breeding of horses and ponies and their management, both during the pregnancy and afterwards.

Foetal Development

As we have already seen, to start the process of growing from egg to foal, fertilization must occur. Fertilization is carried out by a single sperm: from the many millions ejaculated, it takes just one sperm to penetrate the tough outer shell of the egg. This sperm has had to swim all the way to the mare's fallopian tube as this is the only place where normal fertilization occurs. Interestingly, unfertilized eggs never complete the journey to the uterus, but are instead, reabsorbed whilst still in the fallopian tube. Once the sperm has penetrated the egg, the combined genetic material from sperm and egg sets the development wheels in motion and the long process from single cell to foal begins.

The fertilized egg arrives in the mare's uterus about 5–8 days after conception. At this point it has already undergone the first stages of development, the cells rapidly dividing – 2 to 4 to 8 to 16 cells and so on. This is an important fact as it means that should the mare suffer from endometritis the veterinary surgeon can safely treat the uterus with antibiotics without harming the conceptus.

For about 16 days, the little conceptus moves about the uterus and uterine horns, feeding from a small yolk sac. It is thought that the reason that it moves around the uterus in this way is to generate chemical messages to the mare and prompt her body to recognize that she is pregnant. Throughout this time, the conceptus is growing continually and, at about 21 days, the sphere is about 6 cm in diameter with a tiny embryo now visible (by ultrasound) at one end.

At about this time, the conceptus settles in one place in the uterus, normally at the base of one of the uterine horns, and begins the process of implantation. The yolk sac is almost all used up and now membranes that will become the placenta begin to form and attach to the uterine lining of the mare. The placenta and foetal membranes will surround the developing embryo and provide protection and nourishment until it is ready to be born.

The placenta in the horse covers the whole of the surface of the uterus when it is fully formed. (This is not the case in other mammals, such as humans where only a small surface area is covered by the placenta.) This is the main reason why twins are rarely carried to full term in the horse as it means that two placentas will share the uterus. Each placenta will compete for space to attach (Figure 4.4) and, even if one does gain more space than the other, both will be disadvantaged. The equine placenta attaches to the lining of the uterus by means of many microscopic 'button-like' structures called microcotyledons, which actually invade the surface of the uterus. Each of these buttons is composed of tiny finger-like projections that are designed to allow the efficient transfer of nourishment from the mother to the foetus via the mare's bloodstream, as well as the removal of some waste products. When the foal is born, these buttons simply detach from the uterine wall without causing much blood loss.

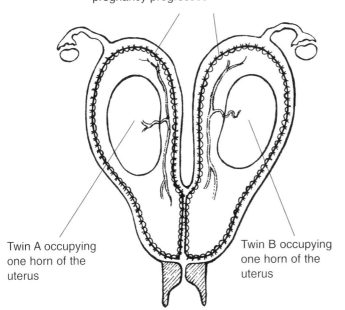

Twin placentas of equal size. In this situation twins may survive but will be compromised as the pregnancy progresses

Twin A occupying one horn of the uterus

Twin B occupying one horn of the uterus

Figure 4.4 *Twin pregnancies commonly compete for space in the mare's uterus and, ultimately, both will be disadvantaged*

The foetal membranes are made up of the allanto-chorion, the amnion and the umbilical cord (see Figure 4.1). Of the three, the allanto-chorion membrane is the nearest to the uterine lining and it develops the finger-like projections, composed of minute blood vessels, which invade the endometrium and make up the microcotyledons. Due to the complex structure of blood vessels, the outer surface of the allanto-chorion is velvety red in appearance, but the inside surface is shiny and white in colour.

The amnion is the membrane that is closest to the foetus; it is opaque white in appearance. Enclosed by the amnion, the foetus is bathed in amniotic fluid; this special fluid is designed to cushion the foetus from the mare's movements – essentially to act both as protection and shock absorber. The allanto-chorion and amnion membranes are richly supplied with blood vessels and the allanto-chorion has many strong veins and arteries that develop during the pregnancy to support the foetus. These two membranes are separated from each other by allantoic fluid, a quantity of liquid composed of waste products from the foetus being the 'water' seen when the mare begins to give birth.

The umbilical cord forms the only attachment between the foetus and the placenta, as well the only attachment between the allanto-chorion and the amnion. It is normally about 70–90 cm long in the Thoroughbred, joins to the foetus at the navel, is normally connected to the placenta at the base of the two uterine horns, carries oxygenated blood to the foetus from the placenta via an umbilical vein and carries deoxygenated blood back to the placenta via two umbilical arteries. A point to remember is that, elsewhere in the body, arteries carry oxygenated blood away from the heart and veins carry deoxygenated blood to the heart – this situation is reversed with the foetus, as its own lungs are not functional. Therefore, deoxygenated blood is carried back to the placenta by the umbilical arteries and oxygenated blood to the foetus by a vein. In addition to the two arteries and the single vein the umbilical cord also includes a tube, the urachus, which transfers urine from the foetal bladder to form the allantoic fluid.

Often, when the placenta and membranes are passed after birth, it is possible to find an additional structure known as the hippomane which varies in size from about 5–14 cm in length and is often brown in colour with a smooth surface. The hippomane is formed from dead cells and other waste products that have, essentially, clumped together in the allantoic fluid.

The Hormones of Pregnancy

Earlier, we looked in detail at the oestrous cycle of the mare and the important role that hormones played in making sure that both mare and stallion had the optimum chance of success in reproduction. So, what happens when the mare is pregnant – does all this activity stop and what are the hormones that help to maintain the pregnancy?

If we think back to the oestrous cycle, we recall that once a follicle ovulates from the ovary, the *corpus luteum*, or yellow body, is formed. This yellow body then begins to produce the hormone progesterone. In a non-pregnant mare, the yellow body would be dissolved by the hormone prostaglandin from the uterus at about 16 days and the cycle would then begin again. Once the mare is pregnant, the cycle in this form halts and hormones are now produced to help maintain the developing pregnancy.

The first stage is to stop the production of prostaglandin and allow the yellow body to continue to secrete progesterone much longer than normal. The ovaries are however not dormant during pregnancy and in fact, follicles are still produced and ovulations occur. Oestrogen, the main hormone of oestrus is also present throughout the pregnancy and is produced not only from the ovaries, but also by the foetus and placenta. There are also certain specialist hormones that are released only during pregnancy and one of these is equine chorionic gonadatrophin (eCG) – which used to be called pregnant mare's serum gonadatrophin, or PMSG. The eCG is secreted by special structures known as endometrial cups that form between the lining of the uterus and the conceptus at about 35 days from conception, when the conceptus first begins to implant. The role of endometrial cups is to help maintain the pregnancy by producing eCG. Once the cups have formed, they will continue to produce eCG until about 90 days. The important fact to remember is that, once formed, the cups will continue to produce the hormone for the same length of time, even if the pregnancy is lost. Testing for this hormone in the mare's blood can therefore produce a false positive result as eCG stimulates the ovaries to continue to produce follicles; the follicles ovulate to keep up the supply of progesterone until about 90–120 days of pregnancy, when the ovaries stop their activity until after the foal is born.

So, where does progesterone come from for the rest of the pregnancy? Progesterone is essential for the whole of the pregnancy and from now to shortly before birth, production is taken over by the placenta and foetus itself. It is not, strictly speaking, the same kind of progesterone as that found in the mare's bloodstream earlier in pregnancy (it is in fact a metabolite of progesterone called progestagen) but it has the same function as progesterone.

It is amazing to think that the tiny embryo has all its organs, albeit in a primitive form, by only 35 days from conception. However, all these organs need to mature and grow, as does the foetus itself to be able to survive independently when it is born. Foals born prematurely struggle to live: although they may have all their organs, most of the vital structures, like the lungs, do not mature to a working form until very late on in the pregnancy.

As the foetus grows in the uterus it takes up more and more room and, from about 80 days, it has grown too big for the horn of the uterus that it had started its development in and spreads to the body of the uterus. As it becomes larger still, only its hindquarters remain in the horn and the majority of the foetus is in the body. When the foetus gets to this size, it cannot turn or move about very much and it normally lies upside down with its back to the mare's belly, or on its side, but in either position its legs, neck and head are flexed. The foetus suspended inside the mare's abdomen, bathed in fluids in the dark and warm uterus does not recognize that it is upside down and so this position is quite natural and probably very comfortable for it – here it remains until it is time to be born.

Pregnancy Diagnosis

Finding out whether a mare is pregnant following mating can be an exciting time. From a management point of view, it is vital to establish that the mare is successfully pregnant with a single foetus as early as possible. In the past, the techniques available to diagnose pregnancy were limited and most required waiting a relatively long time after breeding to be accurate. None of these tests could rule out the presence of more than one pregnancy, or if there was a problem with the growing foetus. Twin foetus' are rarely successful in the horse and most result in early abortion or death shortly after birth (see Chapter 5). In the last 20 years, the use of ultrasound for pregnancy diagnosis has been perhaps the single most important advance in stud management. The technical details of how ultrasound works are complex but, in very simple terms, the scanner produces sound waves at set frequencies that are emitted via a probe. These sound waves will pass through liquids but bounce back from solid structures, producing echoes. The information produced by passing the probe over an area of the body is translated to images that are displayed on a small screen on the front of the scanner. Ultrasound is widely used in human and animal medicine for monitoring pregnancy and as a diagnostic tool for injury and disease investigations.

If the mare is pregnant following mating, she will not return to oestrus and will remain in dioestrus, rejecting the advances of the teaser or stallion. Most stud managers will routinely tease the mares from about 12 days after mating, or 12 days after ovulation if this was confirmed. If the mare has failed to return to oestrus at about 16–18 days after mating, she will be examined by the veterinary surgeon. As mentioned earlier, teasing the mare from 12 days may seem premature, but if she is suffering from uterine inflammation she may come back into oestrus earlier than normal. There are several reasons why the mare may fail to return to oestrus other than because she is pregnant, so this should not be taken as absolute confirmation that the mare is in foal.

It is common practice for most veterinary surgeons to examine the mare's reproductive tract by rectal palpation and with an ultrasound scanner. Examining by rectal palpation at about 18 days from mating, the veterinary surgeon can feel the muscle tone of the uterus, which changes in the pregnant mare and becomes 'turgid' or firm. Examination by speculum shows the cervix is tightly closed to protect the developing pregnancy and, in fact, the cervix becomes further sealed with a mucus plug as the pregnancy progresses. Experienced veterinary surgeons can establish whether the mare is pregnant just from the 'feel' of her uterus, however, this examination cannot always confirm whether there is just a single foetus or what is its state of health. From about 21 days after mating, the pregnancy can normally be felt as a swelling at the base of one of the uterine horns, however, after about 60 days this is more difficult to feel as the swelling has become much larger in size and there is normally a large quantity of fluid in the uterus. After about 120 days, it is possible for the veterinary surgeon to feel the foetus again and from about six months it may be possible to feel it move.

Examination of the reproductive tract with an ultrasound scanner is the most accurate method of diagnosing pregnancy if carried out by an experienced veterinary surgeon, as it provides a clear picture of the condition and stage of development of the foetus. It is routine for large studs to have their mares examined and scanned several times during the early stages of pregnancy. This is because the incidence of early pregnancy loss is quite common in the mare although it may be difficult to find a cause in each case.

The normal pattern for examination and scanning is initially at 15–18 days from ovulation, again at 21–28 days and again at 35–40 days. It is possible to see a pregnancy as early as ten days from ovulation using ultrasound, but most first examinations are carried out five to eight days later.

Scanning the uterus and ovaries is carried out by inserting the probe of the scanner into the mare's rectum. At about 15–18 days from ovulation, the pregnancy, or conceptus, looks like a small black bubble on the scanner screen. This is because it is mainly filled by a yolk sac at this stage and the sound waves pass through the fluid giving a black image. The conceptus is able to move about the uterus at this time so the veterinary surgeon will check the whole uterus thoroughly to ensure that there is only one conceptus. It is normal practice for the mare to be rescanned again a few days later; this is to ensure that the conceptus is continuing to develop and that there is still only one present.

Twin conceptions are not uncommon and may, due to the small size of the conceptus at 15-18 days, be missed on first examination. It is also possible for the mare to have ovulated again on a second follicle a few days after the first and, with a fertile stallion, for this egg to be fertilized as well. Experienced veterinary surgeons are able to identify the presence of a second conceptus and remove it by squeezing it gently through the wall of the uterus. Removing a twin conceptus in this way is a specialized job and only experienced veterinary surgeons are able to carry it out without jeopardizing the other conceptus at the same time. At about 21 days, the embryo can be seen and at about 28 days it is possible to see a heart beat. By 35 to 40 days, the foetus can be clearly identified, together with the developing foetal membranes. (The embryo is correctly termed a foetus from about 35 days.) Experienced veterinary surgeons may also be able to identify the sex of the growing foetus by ultrasound examination at 65–90 days.

In addition to teasing, veterinary examination and ultrasound scanning, blood can also be tested to confirm pregnancy. Blood tests are not performed as routine today as the use of ultrasound has superseded them, but further tests may be useful if the veterinary surgeon is concerned about the development of the pregnancy. By taking blood samples from the mare, the veterinary surgeon will be able to measure the levels of certain pregnancy hormones circulating in the mare's bloodstream.

The hormones measured are progesterone, equine chorionic gonadatrophin (eCG or PMSG) and placental oestrogens. Progesterone is measured from about 18 days after ovulation, eCG appears after about 40 days until about 80–120 days and placental oestrogens are present in blood and urine from about 150–300 days.

Care of the Pregnant Mare

Now that the mare is safely pregnant, we need to look at the best way of managing her to provide optimum care for both the mare and her developing pregnancy. Providing the best care for the pregnant mare ensures that she is able to provide quality nourishment and protection for her foetus throughout the entirety of gestation, as well as providing the best chance for her to produce a strong, healthy foal and recover quickly from giving birth.

Protecting the mare from any unnecessary risks is important throughout the whole of her pregnancy, but this need not mean that she needs to be kept very differently from how she was before. The main risks to her and her foetus are from stress, infection and poor nutrition.

Stress

Stress is an important factor during the early weeks of pregnancy. As discussed earlier, there is a period of some 35–40 days following conception when the little conceptus is mobile inside the mare's uterus and this can be a vulnerable time. Often stallion studs will advise against transporting mares until well after this period, especially if the mare has a history of losing pregnancies. Stress can also affect a mare on her home stud if she is turned out in a large group of mares, exacerbated if the members of the group keep changing. In an ideal situation, pregnant mares should be kept in small settled groups and care taking when reintroducing mares back into the herd. On large studs where there may be a number of pregnant mares, it is common for small groups of mares to be established with similar foaling dates. This means that the mares remain together throughout the whole of their pregnancies and the herd group can be maintained after foaling with the foals also all being of a similar age.

Exercise

Keeping the pregnant mare in work is perfectly acceptable during the first few months and in many cases, it can be beneficial to the mare. Even racehorses can continue in training and race during the first few months of pregnancy, although the Jockey Club does not allow mares to race after 120 days of pregnancy. So the question to work or not to work will very much depend on the individual mare, but by the last third of pregnancy most mares are under enough physical demands from the growing foetus without having to be in work as well.

Lack of exercise can be just as harmful as over-exertion to the pregnant mare, so even if she is not to be ridden, she needs daily exercise such as being turned out to graze in the paddock. As the pregnancy increases in size, the mare may become more reluctant to move about and it is common for minor problems such as stiffness and swelling of the lower leg to become apparent. The sheer increase in weight of the mare may also exacerbate old injuries, especially in the case of mares retired to stud following injury. Exercise improves the circulation to the uterus and, therefore, improves the flow of nutrients to the foetus as well as assisting in the rapid removal of waste products and this is so important because the well-being of the foal is completely dependent on the mare's blood system until it is ready to be born.

Infection

Reducing infection risk to pregnant mares is vital. As in humans, there are certain infections that can cause abortion or damage to the foetus during its development. For this reason pregnant mares should be kept well away from other youngstock as well as from in-training racehorses, competition or performance horses. In Chapter 3, we discussed the importance of preventing infection on a stud and most of the principles remain the same even for mares returning to their home studs. Youngstock, like children going to school for the first time, tend to get everything going as their immature immune systems develop and competing horses are out and about and exposed to a whole range of infections from horses they may come into contact with at racecourses, events, etc.

Nutrition

The pregnant mare does not require any extra feeding during the early months if she is in good health and condition at the start of her pregnancy. As with feeding any horse, quality of the feedstuff is more important than quantity. If the mare is not in work during the early part of her pregnancy, she requires nothing more than a maintenance ration. This should be increased in the last third of her pregnancy as the developing foetus will be growing rapidly in size and the mare will start manufacturing milk. The peak of a broodmare's nutritional needs are actually after the foal is born and during the first few months of its life when all of its needs are met by the mare's milk.

Lactation is discussed later, but the nutritional demands on a mare are greater during this time than those on a horse at peak fitness and at the highest level of competitive work.

As the mare gets closer to her due date, she may become fussy about what she eats and, again due to the sheer size of the pregnancy, be reluctant to eat very much at any one time. Offering a variety of succulent feedstuffs as well as high-quality concentrates can help in ensuring that she is still receiving sufficient nutrients for her needs. In addition, broodmares may also benefit from a suitable vitamin and mineral supplement, although many of the prepared balanced concentrates for broodmares will already have all the required vitamins and minerals added.

Veterinary Examinations

Even though the mare has been diagnosed as being in foal, possibly having had several examinations during the early stages, it is important to have her re-examined for pregnancy in the autumn months. This is of further importance should there be a stud fee to pay at this time – many Thoroughbred studs require payment on 1 October for the stallion's stud fee unless the mare certified barren by 1 October.

However, even if there is no financial urgency to having the mare re-examined, it is still worth doing so from a management point of view. It is not uncommon for mares to be examined as not in foal in the autumn following the breeding season, even when they appear to have had no problem at all in getting pregnant. It is also not uncommon for this fact to have gone completely unnoticed – the foetus is still very small during the early months and can be lost without any outward sign from the mare and, as mares are often out at grass during the late summer months, the slipped pregnancy is commonly quickly removed by animals such as foxes.

Vaccination

A strict policy of vaccination can also help to safeguard pregnant mares from infection. The significant infections that affect studs were discussed in Chapter 3. Of these, EHV is perhaps the most important infection affecting the pregnant mare. Most, if not all, commercial studs require visiting pregnant mares to have been vaccinated against EHV before their arrival at stud and details of the vaccinations to be confirmed in her passport/identification papers signed by a vet. A strict policy of vaccination helps to reduce the risk of infection outbreak not only to the resident horses but also to the visiting mares.

All mares should also be vaccinated in accordance with manufacturers' recommendations for influenza on an annual basis and for tetanus very three years. However, many breeders like to give the mare a booster vaccination for tetanus about 4–6 weeks before she is due to foal to ensure that a good level of protection is transferred to the foal through the mare's milk. There is now vaccination available that combines protection for EHV as well as influenza and tetanus, which may be very useful for breeders. However, it is currently not licenced for viral abortion.

Routine Care

Maintaining a programme of routine care for broodmares is essential. It is surprisingly common for broodmares to be forgotten until they are close to foaling, especially if they are just turned out to grass. The three main procedures that all broodmares will require are farrier attention, regular worming and, annually, their teeth checked.

It is important that broodmares still receive regular farrier attention, particularly as foot problems and old limb injuries can be exacerbated by the increase in size and weight of the mare in the later stages of pregnancy. Most broodmares will have their shoes removed when they are turned out to grass and only have front shoes on if essential. Mares close to foaling should have their shoes removed in all but the most exceptional circumstances so as to avoid risk of injury to the newborn foal. If this is not possible, the mare should have her hooves suitably wrapped to reduce the risk of injury.

Worming is also a routine procedure that should continue throughout pregnancy. All of the modern drugs are safe for use at any stage of pregnancy at the normal dose rate, but veterinary advice should be sought if in doubt. One of the most significant parasites that affect foals, *Strongyloides westeri*, is actually passed to the foal from its mother via her milk. Infection with this particular parasite can cause diarrhoea that, in some cases, can lead to permanent damage of the foal's digestive system. Adult horses have a degree of immunity to this parasite, but it quickly infects foals – in fact, eggs can be found in the faeces of infected foals of just one week old.

Proper management of paddocks and housing used by pregnant mares can help to reduce the risk of parasite infestation – vital to the health of any horse – and relatively easy to achieve. The most important method is to keep paddocks and other grazing areas free from droppings and, although this can be labour intensive, it is worth every ounce of effort. Reducing the numbers of horses grazing an area as well as ensuring that all newly arrived horses are wormed 24–48 hours before being turned out with the resident mares can also be of benefit.

Problems Affecting Pregnancy

Most of the problems that occur during the later stages of pregnancy are minor, such as stiffness and limb inflammation. However, there are occasionally serious problems that do occur which require prompt veterinary attention.

In the last third of pregnancy, the size and weight of the pregnancy may cause the mare some discomfort. This may only be temporary but it often has similar signs to colic. Discomfort can be caused by a variety of factors such as the position of the foal, digestive upsets due to the lack of space in the abdomen or, much more seriously, uterine haemorrhage or torsion. If the mare is showing signs of discomfort, however spasmodic, veterinary attention should be sought as a matter of priority. Another condition associated with late pregnancy and often caused by the increased weight of the pregnancy, is rupture of the pre-pubic tendons – those that support the abdominal muscles and pelvis. This condition requires immediate veterinary attention, but thankfully is very rare.

Vaginal discharge from a pregnant mare should always be brought to the attention of the stud's veterinary surgeon. As mentioned earlier in this chapter, the uterus is sealed off from the outside world by a thick plug of mucus in the cervix. However, if infection has managed to take hold, either in the vagina or by breaching the cervical barrier, it could affect the pregnancy. The veterinary surgeon will need to examine the mare to check the condition of her pregnancy and take a swab of the discharge to be able to identify the infection type and therefore the treatment that is necessary. Occasionally, a discharge is noted in a heavily pregnant mare when she is close to her due date and this may simply be the loss of the mucus plug from the cervix in readiness for the birth of her foal. However, it is always wise to have the mare checked if in any doubt. Any sign of bleeding from the vagina of a heavily pregnant mare should be treated as a veterinary emergency.

As the mare gets closer to the date when her foal will be born, her body begins to prepare itself for the foal's arrival. This includes the production of colostrum, the first milk in the mare's udder. Colostrum contains the antibodies necessary to protect the foal from infection through its first few weeks of life and, as such, it is essential. Some mares begin to produce and actually lose colostrum some days, or more seriously, weeks before giving birth. Individual mares may be prone to producing milk far too early for their foals and do so routinely for each of their pregnancies. Other mares may produce milk early as a sign that there is a problem with their pregnancy and it is often associated with impending abortion. Again, a mare suffering from early milk loss should be brought to the attention of the stud's veterinary surgeon.

5 Abortion

The term abortion is used to describe the loss of pregnancy before 300 days of gestation. The terms early foetal death or resorption are more commonly used to describe the loss of pregnancy in the first 100 days. Stillbirth is the common term to describe the production of a dead foal after 300 days of gestation.

In the previous chapter, we discussed how the foal develops from a tiny conceptus to become a foal ready for the outside world. The foetus may develop all its vital organs early on in the pregnancy, but many of the organs do not mature sufficiently enough to work independently of the placenta until the very last stage of gestation. Foals born prematurely before 300 days of gestation (Figure 5.1) may arrive alive but are unlikely to survive as many of their organs, such as the lungs, are not yet mature. Premature foals born around 300–320 days can survive but they may require round the clock specialist veterinary care, and they may always be compromised by their poor start in life.

Depending on the stage of development and the cause of the abortion, the mare may not give any sign of the impending loss and, indeed, breeders are often completely unaware that it has occurred. It may be that the first indication that the pregnancy has been lost is when the mare returns to oestrus unexpectedly, or when she fails to make the physical changes associated with late pregnancy and the breeder becomes concerned.

Mares generally recover quickly from early abortion, but late abortions can be more complicated. In Chapter 4, we discussed how the foal develops throughout the eleven months of gestation and that the foetus is quite some size in the last third of pregnancy. This factor of size alone can cause complications. If the mare aborts quickly, her body may not have had the time or the physiological 'signals' to prepare for the 'birth'. The pelvic muscles and cervix will not have had time to relax in most cases and damage can occur when the large foetus is expelled. In the next chapter, we will discuss foaling and how the foal moves shortly before or during birth to ensure that it is in the best position to be born: a dead foetus that is being aborted cannot do this.

Abortion can be caused by infectious and non-infectious factors. However, it is vital to treat every abortion, or sign of impending abortion, as infectious until veterinary examination has proved otherwise.

Figure 5.1 *An aborted foetus at about 150 days of gestation*

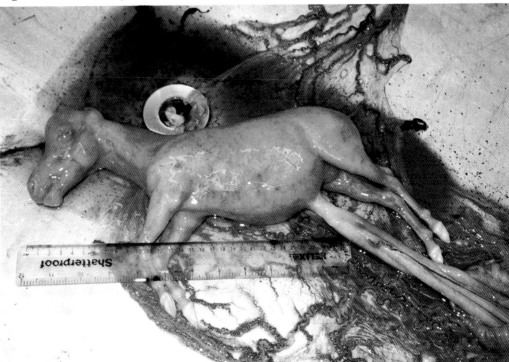

Until the cause of abortion is confirmed, which may take four or five days, complete isolation procedures are essential in the first instance (see later in this chapter).

In many other species, any developmental or genetic abnormality may result in the pregnancy being lost. It is thought that this is Nature's way of ensuring that an abnormality does not survive to reproduce itself. However, some deformities are carried safely to term without any problem such as parrot mouth (Figure 5.2) and cleft palate, etc. Some severe deformities may result in the pregnancy being

Figure 5.2 *An extreme example of a parrot mouth*

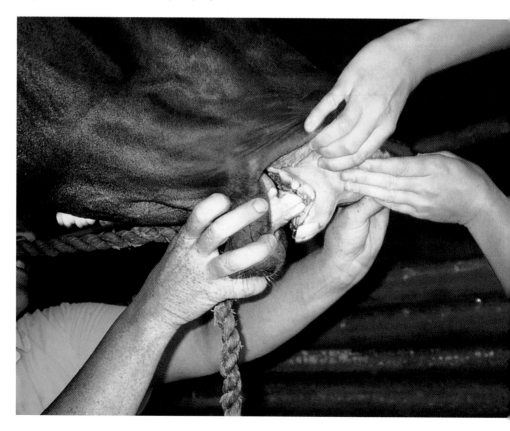

lost as the defect often causes development to cease at some stage and the foetus dies. Other genetic aspects may cause the pregnancy to fail and, as yet, the reasons why are not fully understood. The genetic material from the stallion contained in the sperm is foreign to the mare and should, in any other circumstance, provoke an immune response from the mare. The fact that she carries what is essentially a foreign body without reaction is another marvel of Nature. However, sometimes there is an immune response from the mare and the pregnancy fails – this seems to depend on the individuals that are mated, but the pregnancy will not succeed with the combination of these two individuals whilst the mare's system reacts with an immune response – often a new stallion is chosen for the mare in this scenario.

In addition to the known or likely causes of abortion, a recent (April 2001) outbreak of unexplained pregnancy loss in Kentucky, USA has provoked the need for further research and a careful look at management practices on studfarms. The causal factor for this outbreak of pregnancy losses is, as yet, unknown and it has been called Mare Reproductive Loss Syndrome. Affected mares have aborted at all stages of gestation and live but weak foals have been born to affected mares at term. Disturbingly, there appear to be none of the usual infectious or non-infectious reasons involved in the abortions and neonatal deaths. Research, as already stated, is ongoing but early theories are that the losses may be caused by environmental toxins, such as fungal spores or caterpillar faeces on paddocks and grazing areas and, possibly, trees that the horses have had access to. At the time of writing (2002), this syndrome has not been recorded in the UK.

Preventing abortion is one of the main considerations any manager or breeder takes when deciding on the best care for his/her mares. In some cases, nothing can be done despite the best of care and every precaution being taken. However the following points should be considered when looking after broodmares:

- Broodmares should be examined by a veterinary surgeon prior to breeding to ensure that they are in good reproductive health.
- Stallions used for breeding should be tested free from disease.
- Stud hygiene is of great importance and high standards of stable management should be maintained at all times.

- Good vaccination programmes should be implemented and insisted upon, with veterinary advice if necessary.
- Breeding horses should be receiving a well balanced and good quality diet.
- High levels of paddock management should be practised in studs. Paddocks should be kept as 'clean' as possible and the use of insecticides, herbicides, etc., kept to a minimum if used at all.
- Stress to all breeding stock, particularly pregnant mares at vulnerable times, should be reduced to the minimum wherever possible.
- Mares which have aborted before or have a history of early pregnancy loss should receive particular care and monitoring.

Non-infectious Causes of Abortion

The main non-infectious causes of abortion are twins, hormonal problems, placental problems, management influences and trauma.

Twins

Before the advent of ultrasound scanning, twins were the single most common cause of abortion in mares (Figure 5.3). Twins are rarely carried to full term, many being aborted dead, others being born prematurely and dying shortly afterwards. In twin pregnancies there is also the possible situation where one twin dies in the uterus and becomes mummified – essentially sealed off from the other twin who continues to grow and develop. Rarely is it the case that this surviving twin will have a chance to develop and grow as normal. The first indication of a mummified twin is generally found on examination of the placenta after an apparently normal birth. More commonly, the presence of a mummified twin causes the eventual loss of the remaining twin foetus.

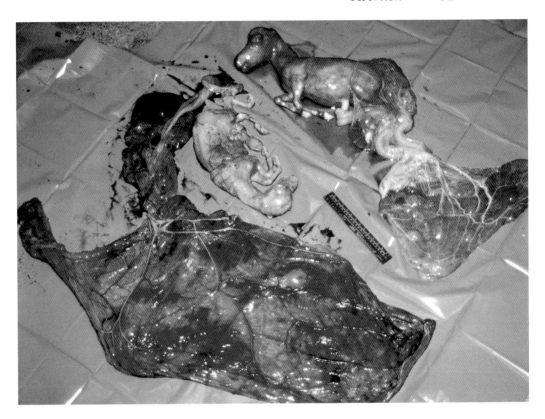

Figure 5.3 *Twins aborted at about 140 days gestation. It is clear that the twin on the left has been suffering for some considerable time by its emaciated condition and the appearance of its placental membranes which have begun to degenerate*

Twin pregnancies are unlikely to succeed because the competition for space and nourishment inside the mare's uterus is too great. We saw in Chapter 4 how the placenta develops: the equine placenta covers the entire surface area of the uterus, so there simply is not enough room for two placentas. As the pregnancies grow in size, the demand for nourishment for each foetus has outgrown what is available and soon one of the twins will suffer whilst the other benefits. Eventually, the smaller foetus will die of malnourishment and its death will usually cause both to be aborted.

Twin conceptions are common in some breeds, notably the Thoroughbred, but this may be due to the increased veterinary surveillance of the mares on Thoroughbred studs as compared to other breed studfarms. It may also be true that the modern selection methods for breeding horses and the veterinary techniques available allowing breeding success with previous problem mares, may increase the number of twin conceptions that are being recorded (see Chapter 4).

Hormonal Problems

A reason sometimes given for why a mare aborts, in the absence of any other obvious reason, is that of hormonal problems. In Chapter 4, we saw how the conceptus is thought to have a role in signalling to the mare that she is now pregnant and should not return to oestrus. Pregnancy loss at an early stage may only be noticed if the mare was examined by ultrasound and then re-examined some days later and found not to be in foal. It may be that this loss is due to the failure of the mare to recognize that she is pregnant and therefore suspicion may point at a breakdown in the signalling. (Early pregnancy loss is also commonly associated with uterine inflammation.) It should be remembered that the conceptus is very small at this stage and its loss is more often termed resorption rather than abortion, as it is uncommon for it to be passed out vaginally before about 40 days of gestation.

The loss of an established pregnancy may be attributed to hormonal failure. From about 140 days of gestation, the foetus itself takes over the hormonal control of the pregnancy and it is reasonable to assume that any error in the development of the foetus may lead to this control being inadequate or incorrect. In Chapter 1, we looked in some detail at the oestrous cycle of the mare and discussed the fine balance between the level of each hormone and the condition of the target organ. It can easily be understood how an imbalance during pregnancy, particularly of the main pregnancy hormone, progesterone, could lead to the death of the foetus and an abortion.

For this reason, many stud managers believe that giving progesterone to the pregnant mare throughout the early stages of pregnancy can help to prevent a loss. Usually this is in the form of a synthetic progesterone preparation such as 'Regumate', given in the mare's feed. Studies have not yet proved whether such supplementation is effective.

Placental Problems

Another causal factor of non-infectious abortion may originate in the placenta. As we have already seen, the placenta is the organ that will nourish and protect the foetus until it is ready to be born and, if its function is compromised for any reason, it is obvious that the foetus will suffer as a consequence.

In the situation that occurs when a mare suffers from chronic endometritis, and is termed a 'problem breeder', the effect of long-term inflammation and/or infection of the uterus is that the delicate lining cells become permanently damaged. This will result in scarring of the uterine wall and loss of the lining cells. The equine placenta completely covers the surface area of the uterus and it is this entire surface area that is necessary to provide the correct flow of nourishment to, and removal of waste products from, the foetus. As with twin conceptions, any compromise in the placental area can result in the malnutrition of the foetus and ultimate loss. Imagine the surface of the chronically affected uterus, pitted with scars that do not allow the formation of the intricate blood vessels to facilitate transfer of nutrients from mare to foetus. Depending on how extensive the area of damage is, the foetus may suffer problems such as failure to thrive and may be born at term underweight or weak. In more serious cases, the remaining unaffected area of the uterus is simply not enough to permit the foetus to grow on and, as the demand becomes too great, the foetus dies triggering the mare to abort.

Indications that all is not well with the uterine lining will be evident on the placenta passed either at a normal birth or with the aborted foetus. The rich red velvety appearance of a normal placenta is marked by areas of pale white tissue where none of the finger-like villi of the blood vessels have been able to develop. Sometimes the pale areas may have a discharge present where the uterine glands have tried to overcompensate for the error.

It is also possible for the foetus to be lost due to the length of the umbilical cord – the lifeline from the placenta and mare to the foetus. The placenta contains major blood vessels to allow oxygen and nutrients to pass from the mare, as well as providing for the removal of waste products from the foetus. In a normal Thoroughbred pregnancy, the cord is in the region of 70–90 cm long, but there are cases where the cord is considerably longer. It may be that this is an inherited trait and research is currently being carried to establish if there is any link. The cord may become entangled around the legs of the foetus as the pregnancy progresses and some degree of twisting is normal. However, in extreme cases the cord becomes so twisted that it cuts off the blood supply to the foetus and does not allow the removal of waste products resulting in the ultimate death of the foetus. The cord may also become compromised for a short time and this may cause it to develop bulges along its length without actually causing harm to the foetus; these bulges are evident on examination after birth. Overlong cords can also cause stress on the foetus due to the extra effort required for normal circulation of blood from placenta to foetus and back to placenta again.

Failure of the placenta to properly develop and implant is a common reason put forward for pregnancy loss in the first month of gestation. During this time, of course, the foetus is moving free within the uterus and is particularly vulnerable until it is properly attached.

Management Influences

How mares are managed, especially when they are pregnant, may be of relevance in seeking the cause of abortion in a mare. Some mares are temperamentally more prone to suffer from the effect of stress and may be more likely to lose their pregnancies. However, in most cases mares seem remarkably resilient to any amount of stress during pregnancy and remain unaffected by quite dramatic events such as being put through a sales ring, low-flying aircraft close to their paddocks, travelling by horsebox, making a sea crossing, or being flown abroad. There are certain stages in the pregnancy when the mare may be more susceptible to stress, such as during the early weeks before the placenta has properly implanted and at the later stages when she is close to term. In most cases, even without direct evidence of stress inducing abortion, it is wise to keep pregnant mares as settled as possible.

Trauma

Amazingly, the foetus is very well cushioned against all but the most direct physical traumas, such as kicks, falls, etc. Surrounded in fluids and suspended within the mare's abdomen, the foetus is protected and, in the wild state, the heavily pregnant mare would need to be able to flee from prey and the way she carries her pregnancy is designed so as not to hinder her survival.

If the mare's life is threatened by the trauma then the pregnancy may be aborted as a direct result, but often the pregnancy continues as normal even after extreme events such as the mare requiring surgery for colic.

Infectious Causes of Abortion

Infections of the placenta may occur as a direct result of a pathogen that was present in the uterus at the time of conception, or they may occur by entry through a compromised cervix during the pregnancy. Significantly, infection can also be passed via the mare's bloodstream to the foetus and cause abortion.

Bacterial and Fungal Infections

The entry of bacterial and fungal infections to the uterus is generally via the cervix and may happen at conception. It can also occur later in gestation if the cervical seal is insufficient. In Chapter 3 we discussed conformation of the genital tract of the mare and how the various seals of the vulva, vestibule and cervix act to prevent the entry of airborne pathogens. In most cases, if infection occurs before or during mating the pregnancy will be lost very early on as inflammation of the uterine lining will not permit normal placental and foetal development.

Indications of bacterial or fungal infection may be in the form of a discharge from the mare's vulva. The bacteria or fungi invade the placenta around the cervical area and infection will usually extend until it is so great that the foetus dies and is aborted. If the infection remains localized, the pregnancy may go to term and the foal will be born normally. However, the foal may be small and undernourished and the placenta will display signs of infection.

The most common bacteria associated with abortion are *Streptococcus*, *E.Coli*, *Staphylococci* and *Salmonella*, although *Salmonella* infections are now rare in the UK.

Viral Infections

The main viral infections associated with abortion are Equine Herpes Virus, Type 1 (EHV–1) and Equine Viral Arteritis (EVA). We have already briefly mentioned these infections in Chapter 3, but such is their significance as a cause of abortion we should rightly discuss both infections in further detail in this chapter also.

Equine Herpes Virus

EHV–1 is a very common virus – and also very contagious. It causes respiratory infection, abortion and in some cases, paralysis. All types and ages of horses can be affected with EHV–1, but pregnant mares are the most vulnerable to infection and this is the main reason why they should be kept away from other horse groups, particularly youngstock. EHV–1 is very commonly spread via the respiratory route, for example by contact with nasal discharge from an infected horse, or from coughing, sneezing, etc. The virus causes a mild respiratory disease in weanlings and yearlings, normally in the autumn and winter months. The signs are usually mild fever, coughing and nasal discharge. EHV–1 can also cause neurological disease in the horse and this can be devastating.

The first case of neurological disease in the UK was seen in 1966 and pregnant or lactating mares seem prone to developing this complication following EHV–1 infection. Affected horses lose coordination and develop signs of disease similar to that of 'Wobbler' syndrome. Some of these horses lie down and are unable to stand again – the future for these horses is very grave. The birth membranes and fluids of an aborted infected foetus are also loaded with the virus and are a prime source of infection to other mares, etc. Sick or weak foals born alive but with EHV infection are also a risk to all other pregnant mares. Infection with this virus can cause mares to abort without any warning and often all contact mares in the group will abort, giving rise to the term 'abortion storms'. Testing for EHV can be done by taking blood samples and throat swabs in the case of the respiratory or paralytic form and by tissue culture from aborted foetuses and membranes. 'Storms' are far less likely to occur in vaccinated mares.

EHV, like many other viruses, is remarkably adept at surviving outside the horse on fencing, stabling, etc., for several weeks, so indirect infection some time later can also occur. One of the other sinister aspects of EHV infection is that sometimes an infected horse can act as a carrier, showing no outward sign of the disease but transmitting it to all that he/she comes into contact with.

Preventing further infection is paramount and any aborted foetus, together with any birth membranes and fluids should always be sent to a recognized laboratory for post mortem examination. Until infection has been ruled out as a cause of abortion, strict isolation procedures should always be followed. It cannot be stressed often enough that every abortion or death of a weak, sick foal shortly after birth should always be treated as infectious until proved otherwise. If an abortion occurs it is important to inform all owners of mares that may have been in contact with the infected mare and then left the stud so that the owners can make adequate precautions at their own premises.

So how does the virus cause abortion? EHV can pass from the mare to the foetus via the placenta, infecting the foetus and causing it to die. The virus can also cause inflammation of the placental arteries, which will then cause contractions of the uterus resulting in the typically sudden abortion. In these circumstances the foetus does not look sick as it may not have had a chance to become fully infected before the abortion occurred. This is yet another reason why all abortions should be treated as infectious in the first instance.

When EHV was first recognized in the 1930s, it was thought that only abortions occurring between the seventh and nineth months of pregnancy were significant. Today we know that infected mares can abort as early as four months gestation or go on to produce a sick infected foal at term. Mares that become infected may not abort immediately; it is known that abortion can occur anything from a few weeks to some months after the mare was first in contact with the virus.

Stud hygiene is most important in reducing the risk of EHV–1 infection as the virus is vulnerable to disinfectant and heat. Steam cleaning and disinfecting stabling, especially foaling boxes, should be part of every stud's normal routine. The risk of introducing infection onto a stud can also be dramatically reduced by having a strict policy of only accepting vaccinated mares to foal and by ensuring that all the resident pregnant mares are also vaccinated.

Equine Viral Arteritis

Like EHV, the viral disease EVA is very contagious. It is known all over the world and is endemic as close to the UK as mainland Europe. Cases of EVA are regularly seen in the UK in the non-Thoroughbred horse population and this is probably due to the fact that not all non-Thoroughbred studs insist on horses being tested prior to arrival or prior to use as breeding horses. In 2001, cases of EVA were reported in Thoroughbred horses in Germany and France. This resulted in a further tightening of the UK testing procedures for all Thoroughbreds originating from those two countries and for those mares who had visited stallions there in the previous year.

Mares become infected either by mating with an infected stallion, being inseminated with infected semen, by droplet transmission from coughing and sneezing, or by coming into contact with aborted foetuses and membranes. Stallions are a particular source of infection and, as already indicated, the stallion does not actually have to mate with the mare in the traditional manner to pass on the virus, as it is readily transmitted in semen used for insemination. A sinister aspect of EVA is its ability to produce carriers in stallions. Although not all infected stallions become carriers, this outcome is possible and all infected stallions should therefore be tested in accordance with the veterinary surgeon's guidelines. A stallion that is found to be a carrier normally has two possible futures: castration, followed by a further period of isolation, or euthanasia. Infected mares will pass the infection via droplet transmission – the carrier state is not yet known in mares. EVA, as is common with most viruses, can survive being chilled or frozen, as it is resistant to climate changes; therefore, the methods for storing semen for insemination will not affect it.

One of the other main factors that makes EVA infection so significant is that not all horses show signs of infection and the disease can be hard to diagnose before it has done the damage. Those horses that do show signs of the virus may have a fever, be lethargic, exhibit swelling of the lower limbs/sheath/mammary glands, as well as exhibiting swelling around the eyelid and eye socket with conjunctivitis (inflammation of the eye membrane) also being commonly seen. Infected pregnant mares may abort.

Establishing that a horse is infected with the EVA virus is of great importance and early diagnosis can help to reduce the risk of a large number of horses becoming infected. The virus is normally identified by taking blood samples from suspected horses as well as swabs from the back of the nasal passages (nasal pharyngeal). If a mare has aborted, the virus can be readily isolated from the foetus and membranes.

Testing for infection should be carried out on every horse coming onto the stud and repeated each year (see Chapter 3). If all horses are tested in line with the recommended guidelines, including those used just for insemination programmes, it will dramatically reduce the risk of accidental infection. There have been cases where infection has occurred through a foster mare arriving from outside to foster an orphan foal and, because time is of the essence with fostering, she has not been adequately screened before arrival or isolated with the orphan on arrival.

Vaccination against EVA is available under licence in the UK.

However, to date (2002) full research trials have not been completed as to its efficiency. Vaccination is most commonly used only for stallions on the larger commercial studs.

Isolation Procedures

Most studs have some facility for isolation – it does not need to be on a grand scale. However, facilities for complete isolation should be considered essential for studs that regularly board horses from overseas, or from sales.

Essentially, an isolation yard should be located away from the main breeding and foaling areas of the stud. The structure of the stabling should be such that complete disinfection and cleaning can be ensured. The yard should be at least 100 m from any other horses and no contact, either directly or indirectly with horses, should be possible. The yard should be totally self-contained with its own feed and forage stores and muck heap. The staff that attend to the horses in isolation should be separate from those who attend to the other horses if at all possible. If this is not practical then the isolation horses should be done after the healthy horses each day and the staff must maintain extremely high levels of hygiene. Footbaths, overalls and protective footwear should be kept for the isolation yard only. The yard should have its own stable tools, tack and grooming equipment, made of materials that can be cleaned completely and easily.

Keeping horses in such high levels of isolation is labour intensive and time-consuming. However, it is the only way to reduce further risk of infection and any time spent properly in isolation reduces the long-term time and labour implications that would occur if any outbreak affected the whole stud.

Transporting horses may increase the risk of infection of some diseases. When transported, horses may become stressed and more susceptible to infection. If the transport used is not cleaned thoroughly and routinely with the correct disinfectants, it is possible that viruses and some bacteria can remain in the environment for some considerable time.

Associations such as the Thoroughbred Breeders' Association can advise on insurance policies to protect against losses should a healthy mare be forced to remain on an infected stud. The use of this type of insurance assists with the financial costs or losses incurred by a mare owner should an outbreak occur on a stud that the mare is visiting and is commonly called 'Lock-in' insurance.

6 Foaling and the Newborn Foal

The birth of a foal is always a miraculous event regardless if it is the first seen or the thousandth. In almost all cases, equine birth is a straightforward process with few complications; however, if problems do occur they can become life threatening very quickly. Mares vary greatly in the signs they give prior to foaling, so it is important for anyone involved in their care to understand what is normal behaviour and what is not throughout the whole process. The normal length of gestation in the horse is between 320–360 days, although there is wide variation between individuals.

Monitoring Foaling Mares

What specifically starts the foaling process is still not completely understood, but it is known that the foetus, as well as the mare, plays an important part in the complex production of hormones that initiates the preparatory processes prior to birth. Most foalings occur at night, generally in the very early hours. This is thought to be due to an instinctive need for the mare to give birth at a time when she is least at risk from the majority of her predators. Some mares, however, appear able to 'hang on' for a short time if they are disturbed by over fussy attendants or other events. This fact is confirmed every year by frustrated and sleep-deprived owners who have sat up with their mares religiously every night only to have them produce unattended when least expected.

Although mares do foal successfully alone, it is wise for an attendant to be present for the birth if at all possible – and essential if the mare has a history of difficult births. Large studs that have many mares foaling each year, normally have a full time night watchman who may or may not be responsible for the care of the mare and foal during the actual birth. On many Thoroughbred studs, the night watchman's role is literally just to watch the mares and then to call on an experienced person, usually the Stud Groom, who attends to the actual birth. Smaller studs may not be able to justify the cost of another member of staff for the season and may work a rota system with the existing staff or use some other means of monitoring the pregnant mares. There are several varieties of monitoring system available; these range from electronic foaling alarms that the mare wears fitted to a roller, to kits for testing the milk secretions to predict the imminence of birth, and sophisticated closed-circuit cameras.

Most of the commercial electronic foaling alarms work using a small radio transmitter fitted to a roller that the mare wears. If the mare should lie down flat out it will trigger the alarm and an audible signal is sent to the attendant's handset. The radio transmitters have quite a good range and in most cases would allow the attendant to be based in a nearby house or sitting up room. Some more complicated alarms also monitor body temperature of the mare and are triggered if she should start to sweat up – a common precursor to foaling. Some breeders swear by these electronic alarms, but in the author's experience some mares can foal completely without triggering the alarm at all!

Closed-circuit cameras are an excellent addition to the monitoring equipment for any stud with pregnant mares. Modern cameras are very small and can be sited in the stable with little effort. Most of the cameras have special lenses which enable a clear view of the whole of the foaling box with no blind spots. Infra-red lights are also commonly used so that the mare does not have to be disturbed by bright lights to be seen clearly. The monitors for the cameras can be set up in a separate place, such as the main house or in a designated sitting up room – most monitors can be set to observe several mares at any one time, or fixed to view only one mare. The main advantage with closed-circuit cameras is that the mares are unaware that they are being observed (in fact, the whole foaling process can be observed from outside the mare's box) but attendants are on hand if the mare should require assistance. However, cameras alone will not alert anyone to the fact that the mare has started foaling so there still needs to be a member of staff watching the monitors throughout the night.

Predicting exactly when a mare is likely to foal is something that not even very experienced stud staff can get right every time. This fact alone makes looking after heavily pregnant mares very exhausting and frustrating. In addition to the methods of monitoring already mentioned, there are also kits available to test the early milk secretions present in the mare's udder about a week from foaling – the milk which will become colostrum – as she gets closer to foaling. Colostrum is the first milk a mare produces and it contains all the essential antibodies necessary to protect the newborn foal from infection. By sampling the early pre-colostrol secretions and using a stick test from the kit, similar to that used for testing urine, a percentage chance of the mare foaling within a 24-hour period can be established. The kits work by measuring the levels of certain electrolyte elements in the mare's milk and, because some levels will dramatically increase and some decrease within a few days and hours of birth, a fairly accurate result can be obtained. The main disadvantages with this kit is that not all mares have milk secretions present and available for sampling prior to foaling and the kits themselves can be quite expensive.

Foaling Equipment

The equipment kept for foaling will depend on the type of stud, the number of mares to be foaled each year and the level of experience of the stud staff. It is important that the foaling kit is kept just for foaling and can be easily cleaned.

The basic kit should include the following:

- Tail bandages – ideally disposable.
- Two buckets – ideally of a different colour to those used for feeds, etc.
- Bin liners for afterbirth, etc.
- Paper towels – ideally of veterinary grade on a roll for ease of use.
- Disposable gloves.
- Sharp surgical scissors.
- Antiseptic or iodine for navel dressing.
- Foal enemas.
- Protective, washable overtrousers and overshirt for attendant.
- Wellington boots, kept for foaling only, or disposable boot covers.

Large studs may keep other items in their foaling kit such as oxygen, obstetrical ropes, sedatives and antibiotics. These items must be used by an experienced person and/or under the guidelines of the stud's veterinary surgeon. The amount of specialized equipment required will depend also on the number of mares foaled each year and how far away the nearest veterinary surgeon is. For example, in areas such as Newmarket, where there is a concentration of Thoroughbred studs and racing stables, veterinary attendance can almost be guaranteed within 15 minutes of call out – but in isolated areas such specialized equipment in capable hands can save lives.

Preparation for Foaling

As the mare comes closer to her due date, physical changes will occur which indicate her readiness for the birth. These changes include development of the udder (mammary glands), which enlarges as it starts to contain an increasing amount of milk. Some mares show signs of udder development quite early on and some leave it until the very last minute. First foaling mares, also termed maiden foalers, may not show much in the way of udder development until close to the actual birth.

As the udder fills with the first milk in readiness for the foal's birth, some of the milk may drip from the end of the teats and dry to form waxy looking beads. This is commonly called 'waxing up' and is a sign, albeit a bit unreliable, that foaling is likely within 24–48 hours.

Some mares may run milk – literally have milk running out from the teats – for several days prior to actual foaling. It is not completely understood why some mares do this as a matter of routine and others never do, but it is an important point to note. Loss of the first milk, the colostrum, can be a problem as it contains the protective antibodies the foal needs to give it some resistance against infection. Larger studs may keep a store of spare colostrum deep frozen to supplement those foals whose mothers have run milk prior to foaling, or for those who for some other reason have failed to receive sufficient amounts at birth. This colostrum is taken in small quantities (280 ml – equivalent to a standard full baby's bottle) from suitable mares that have just foaled and have a plentiful supply. Normally, colostrum is only stored from mares that have a known breeding history, such as perhaps from the studfarm's own mares. Although it will be immediately frozen, if, following an IgG test on the donor mare's foal, the colostrum is shown to be of poor quality it will not be kept. Powdered artificial colostrum is available from veterinary practices in an emergency but frozen donor colostrum is the ideal.

Colostrum is usually kept in babies' bottles and can be stored in the freezer for some considerable time. When a bottle is required, it should be defrosted gradually in a bucket of warm water until the required temperature is reached. Care should be taken when defrosting frozen colostrum as too much heat can cause the delicate antibody protein to be damaged. Defrosting in a microwave oven or using boiling water is therefore unsuitable.

Another change that occurs in the last weeks before foaling, is that the mare's pelvic area begins to relax. The strong muscles that surround and support the pelvis must relax to ease the passage of the foal through the birth canal. In some mares, this relaxation is very noticeable and they begin to look in very poor condition across their hindquarters, developing hollows on either side of the top of the tail.

In other mares, the change may not be so visible. As with the pelvic muscles, the vulva also begins to relax and lengthen during the last few days prior to foaling.

These are the main external physical changes that the mare is close to foaling, but her behaviour may also change, sometimes quite noticeably. As she gets closer to the birth process beginning, the mare becomes increasingly restless and solitary. She may take herself off to a corner of the field and want to be left alone. In the stable she may become restless, swishing her tail, kicking at or looking round at her flanks – these signs may be similar to that of colic but at this stage, if the mare is passing normal droppings and urinating, it is more likely to be an indication of the start of labour.

Some mares may also begin to sweat as they get close to the start of labour. This type of sweating is quite distinct from that caused by exercise as it is more literally 'steaming up' and can be sudden in onset. Some mares may sweat up and become very restless on two or three occasions over the course of a few days prior to actually giving birth. These 'false alarms' are more frustrating than anything and in themselves, are not necessarily a sign that something is wrong, but advice from the stud's veterinary surgeon should be sought if there is any element of doubt.

The answer to the question of when to move the mare to a foaling box will depend on the facilities available and the normal procedure of each particular stud. Some studs have sufficient foaling boxes to enable the mare to be moved onto the labour ward, as it were, some three weeks prior to foaling. This is the ideal in many respects as it allows the mare the chance to settle in to her new environment and routine, as well as giving her a chance to develop immunity to any environmental pathogens present that she has not encountered before. This immunity will then be passed on to her newborn foal through the colostrum.

However, some studs, in contrast, move their mares to the foaling box literally as the first signs of foaling are apparent and move her back again the very next morning if all is well. Studs who use this method report that it works well, but it would probably only be of use with an established closed herd where the infection risk was minimal.

In the natural state, mares foal very successfully outdoors. In the

UK, the climate does not predispose towards outdoor foaling for much of the year for the more sensitive breeds. However, for the native breeds and for those mares due in the late spring/early summer months, there is no reason why foaling outside should be considered a problem. The most important considerations with foaling mares outside are safety and hygiene for both mare and foal. The paddock used for the foaling mare should be as flat as possible, with a minimum of natural hazards such as ditches and ponds, close to the stud buildings in the event of a problem and not too large to avoid having to spend time searching for the mare at night. In countries such as Australia, even Thoroughbred studs have floodlit foaling paddocks that are used as routine rather than foaling boxes. (Foaling boxes are discussed in more detail in Chapter 8.)

Normal Foaling

The normal events of giving birth are easy to recognize once the inner physical activity of birth is also understood. The emphasis for anyone involved with foaling mares is an understanding of what is normal, so that any variation can be quickly identified and assistance sought for the mare as required.

The whole process of giving birth is commonly split up into three stages.

First Stage

This stage can be thought of as both preparation and positioning.

Preparation

The key word for both mare and soon to be newborn foal is preparation. The mare's genital tract must become ready for the passage of the foal and some major changes need to happen before this can occur. As discussed earlier, the cervix has remained tightly shut throughout pregnancy, protecting the foetus and the placenta from the challenges of the outside world. Now the job has been done and the foal is ready to pass through this small muscular seal to be born. The pelvic muscles relax and the cervix must also relax – dilating enormously to allow the safe passage of the foal. Throughout the first stage of labour, the cervix dilates, the vagina becomes lubricated with mucus, the vulval lips lengthen and engorge to aid accommodation of the foal as it completes its journey to the outside world. If the mare has had a Caslick operation (had her vulva stitched), an episiotomy must be performed. This 'opening up' refers to the stitched area of the vulva being cut with scissors to ensure there is room for the foal to pass through as normal. It is vital that a stitched mare is opened prior to foaling to ensure that she does not become badly torn during foaling. Permanent, sometimes irreparable damage can be done if a stitched mare is left to foal without this procedure being carried out; often it is simply not done because the mare foaled unexpectedly. Mares that suffer vulval tears during foaling tend to develop scar tissue which is thick and fibrous, not the elastic smooth tissue that forms such an excellent vulval seal. After a Caslick has been performed, the stitched area is somewhat stronger than the normal tissue and it will not always be the first area to tear. Sometimes, mares display extensive lacerations the entire length of the vulva and up to the anus. All such tears will require repairing before these mares can be considered for mating again – indeed, severe tears may mean that they cannot be mated again that year. Some studs arrange for their veterinary surgeon to 'open up' pregnant mares using a local anaesthetic a week or so before they are due to foal, other studs will wait until the waters have broken before carrying this out. If there is any risk of the mare foaling unattended, she should be opened as soon as she comes close to her due date.

Positioning

In Chapter 4, the position of the foetus throughout pregnancy was discussed, in that it remains lying upside down, with its back pointing to the ground and its legs facing towards the mare's spine. This is the ideal position for growth and development during the later stages of gestation, but is not the optimum position for being born. During the first stage, the foal begins to move into the correct position for birth, which is with its head and neck and front legs extended (Figure 6.1); essentially, the foal looks as if to dive out of the mare's pelvis as it is born.

The foal is stimulated to move by the start of uterine contractions and cervix dilation which cause the mare to start to sweat up, either a patchy or extensive steaming sweat. She will also show some amount of pain to a greater or lesser degree. There is no doubt that the mare feels considerable pain throughout the process of birth and this is due to the physical changes that are occurring to her body in order to aid the delivery of the foal. She will often pace her box, dig at her bedding and lie down only to get up again immediately, she may even roll and show signs of colic. All of these behaviours are considered normal during the first stage of labour but if this stage seems unduly prolonged or exaggerated, veterinary advice should be sought. The uterus will continue to contract and the cervix dilates. Because of the pressure from contractions and the increase in the size of the opening of the cervix, the allanto-chorion bulges through. The area of the membrane that has lain next to the cervix throughout pregnancy, called the placental pole, begins to thin and will rupture causing release of the allantoic fluid – what is know as 'breaking water'. This event marks the start of the second stage of labour.

Second Stage

Once the 'waters' have broken, the foal must be born. In normal circumstances, the second stage lasts about 25 minutes from the waters breaking to the birth of the foal (see Figures 6.2–6.6).

Once the placenta has ruptured and the allantoic fluid is released, the next event will be the appearance of the amnion, the inner white membrane that surrounds the foal, at the vulval lips. It is often possible to see the foal's forelegs as the amnion appears for the first time. An experienced handler will check that the foal is correctly positioned at this time by gently inserting a gloved hand into the mare's vagina and ensuring that both the foal's forelegs and muzzle are present. The handler will also check that the amnion appears normal and does not

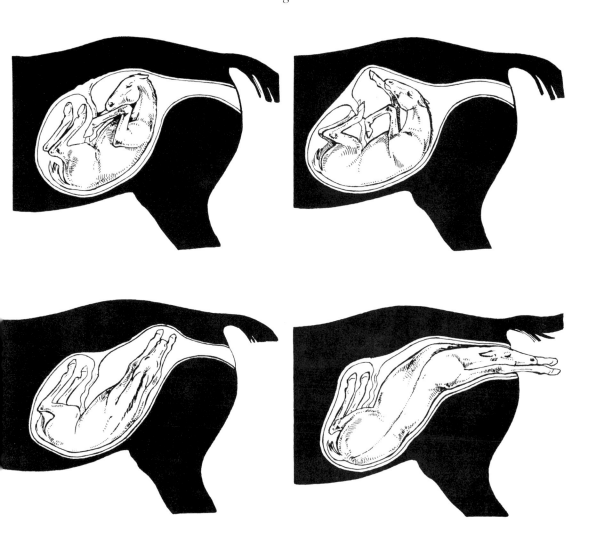

Figure 6.1 *In readiness for birth* (reading from left to right), *the foal repositions itself* (top, left), *rotates its body* (top, right) *and extends its neck and forelegs* (bottom, left). *It then moves into an upright 'diving' position* (bottom, right) *in preparation for birth*

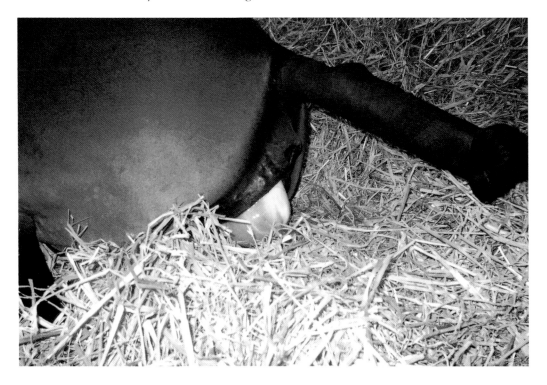

Figure 6.2 *The amnion appears at the vulval lips at the start of the second stage of foaling*

contain stained fluid, as this may indicate that the foal has been distressed either before or during this stage of labour. Apart from these checks, if all is well, the mare should be left alone as much as possible. Ideally, either she should be observed on camera or from just outside the foaling box so that disturbance can be kept to a bare minimum.

The foal's forelegs appear first, often one slightly behind the other, as this helps to accommodate the broad width of the shoulders through the mare's pelvic ring. The foal's head is normally resting on its knees. The mare may lie down and get up again once or twice once the waters have broken, but under normal circumstances, she will stay lying down once the foal's legs appear at the vulval lips. It may be that by getting up, walking around and then lying down again, that

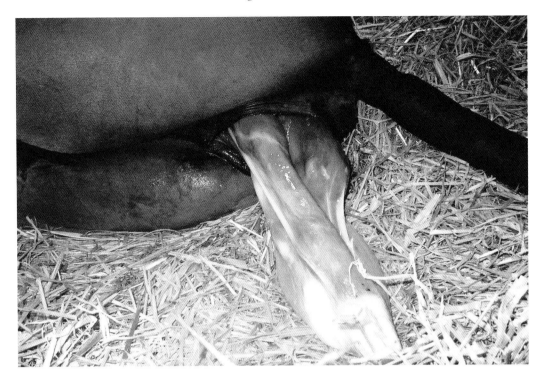

Figure 6.3 *The foal's front legs and muzzle can be clearly seen as the second stage progresses*

the mare is helping the foal to come into the correct position for birth. The contractions continue throughout the second stage until the foal is born. Once the waters have broken, the mare will display obvious physical effort, straining to deliver the foal although contractions are involuntary. By lying down, she is able to fix her diaphragm and use the strong abdominal muscles to literally push the foal out through the birth canal. It is a common fallacy that birth is about pulling the foal out – natural labour is all about pushing.

Once the forelegs and head have passed through the vulva, the remainder of the foal is born quite quickly. It is normal for the foal's hind legs to remain in the mare's vagina for some time after birth and it may be that this encourages the mare to remain lying down and rest.

Figure 6.4 *The foal is still enclosed in the amniotic membrane. Its shoulders have just passed through the mare's pelvis. From this point on, the rest of the foal will be delivered quickly*

If the birth has been normal up to this stage, the mare and foal should be left alone for as long as possible to encourage them to rest and for bonding to begin. The foal may snort and sneeze to clear its nasal passages and within a few minutes will raise its head and move to right itself onto its chest or brisket – it may then rest in this position for a while. Foals often shiver and shake quite alarmingly at this time. This is due to fact that they are trying to regulate their body temperature for the first time. The suck reflex is the next behaviour to be displayed

Figure 6.5 *The amniotic membrane is often broken by the foal's early movements*

and the foal will often protrude its tongue just a little and make sucking movements in the air. The instinctive urge to stand cannot be overcome and the foal will start to move its front legs first in an effort to gain the strength and coordination to stand.

Once the mare stands, an attendant will usually tie up the protruding placenta with string to avoid it being stood on by the mare until it is passed completely. This is a good time to perform brief health checks of both mare and foal to ensure that all is well, before leaving

Figure 6.6 *The hind legs of the foal commonly remain in the birth canal for some time after birth which encourages the mare to remain lying down and rest*

them alone again. Some studs will also offer the mare a warm, wet mash type feed at this time and a drink. The mare may start to eat her bed and seems to have a need for roughage so mash feeds are normally well received. The foal may also be routinely given an enema at this time to help it pass the meconium. Meconium is the first droppings the foal will pass and consists of pellets of green/black, sticky, tar-like substance. Occasionally, these faecal pellets become lodged and the foal shows signs of colic and abdominal pain. Most modern enemas used on studs are of the type used in human medicine and are effective as well as easy to administer.

A new procedure performed on some Thoroughbred studs is the

washing of the mare's legs, belly and quarters with warm water and mild disinfectant immediately after foaling. It is now thought that many of the early infections that foals are prone to may be transmitted during the first few hours of birth when the foal is searching for the udder and sucking at other areas of the mare's body. Foal's instinctively search for the udder in dark places and this is why newborns will often also suck at stable walls painted black. Washing off the mare with a disinfectant solution is thought to reduce the risk of accidental transfer of a pathogen when the foal is at its most vulnerable. However, as already mentioned, the most important role of any foaling attendant is to keep disturbance to an absolute minimum, so the washing of the mare should be carried out as quickly as possible and not at all if the mare is likely to be upset by it.

Bonding between the mare and foal will begin in earnest as soon as the foal is delivered and the mare is able to see it. She will show a great interest in the foal, responding to its movements by whickering and licking. The smell of the birth fluids is also known to stimulate the mare's maternal recognition of her foal. Some stud grooms like to move the foal close to the mare's head so that she can reach it without having to get up immediately but moving the foal may also cause the mare to jump to her feet straight away.

Third Stage

The third stage of labour is completed when the placenta or afterbirth has been passed completely by the mare (Figures 6.7 and 6.8). At the end of the second stage of labour, the placenta is still inside the mare's uterus and the foal remains attached via the umbilical cord. The cord will break quite naturally at a point about 4–5 cm from the foal's navel, this will occur either when the mare gets up or when the foal begins to make its first movement towards standing. It is important to allow the cord to break naturally and only under veterinary instruction should it ever be cut as blood vessels 'seal' less well when cut. There will still be a considerable amount of blood circulating in the placenta that needs to pass through the cord to the foal and, if it is severed, this blood will be lost. Some studs like to dress each foal's navel

Figure 6.7
*The placenta is
expelled shortly
after birth,
during the third
stage of foaling*

Figure 6.8 *The placenta should be thoroughly checked once it is passed to ensure that it is complete. This photograph shows a placenta with the red velvety side outermost; this is the side that would have been attached to the lining of the mare's uterus*

with either antiseptic powder/spray or iodine as soon as the cord breaks, but it will seal itself naturally so this is not strictly necessary.

The placenta begins to separate from its attachment to the wall of the uterus as soon as delivery is underway and has usually come away completely about half an hour after the foal is born. The mare may exhibit some after pain as she strains to pass the placenta, but many mares expel it with no indication of pain just a slight straining. If the mare shows signs of prolonged pain and/or sweating veterinary assistance should be sought immediately as these signs could indicate a complication such as uterine haemorrhage.

If the placenta is not passed it can cause serious problems for the mare and veterinary advice should be sought if it has not come away within ten hours of birth. Never try to pull the placenta out. The placenta must always be checked carefully to ensure that it appears normal and that it is complete with no pieces missing. Again, if there is any doubt veterinary advice should be sought.

Continued observation of the pair is important from some point outside the box until the mare has expelled the placenta, the foal is standing and has had its first drink. It is useful to record the time when these events occur for future reference.

The newborn's efforts to stand can be immensely frustrating to watch as the foal may fall, get tangled in its own legs and generally wear itself out in what seems to be fruitless exertion. However, unless there is a problem the foal will stand successfully within about 1–2 hours of being born. Once standing it will actively seek out the udder for its first feed and, under normal circumstances, the foal should have sucked within two hours of birth. Intervention by handlers in an effort to assist the foal to stand or suck tends to have the opposite effect so, unless there is a good reason for assisting, the foal is best left to learn alone. The one exception to this statement is with first foaling mares who may be ticklish and unsure of what their foal is doing in its search for the udder. It is often helpful to both mare and foal if the mare is held until the foal has had a drink and in doing so released the tension in the mare's udder.

The colostrum or first milk that the mare produces is rich in antibodies (IgG) to enable the foal to have some resistance against disease until it can develop its own. We have already discussed the problem of mares that run their milk early before foaling and storing donor colostrum for later use in such cases. It is important that the foal receives the colostrum within the first few hours of birth as the lining of its intestine changes after about 12–24 hours and the antibodies cannot be absorbed. It is common practice on Thoroughbred studs for the foal to have a blood sample taken when it is about 24–48 hours old to measure the IgG levels present. Sometimes, even if the foal received good colostrum at the correct time, it may have failed to acquire the antibodies – this is called failure of passive transfer. As any foal can be affected by this, every foal should be tested as routine – without adequate antibody protection the foal will have no resistance to disease. For those foals with a low IgG level, the veterinary surgeon may advise an intravenous serum transfusion and/or a course of antibiotics.

Foaling Problems

Difficulty during foaling is often called dystocia. Dealing with problems during birth is beyond the scope of this book, but it is important for everyone involved in the care of foaling mares to be able to recognize an abnormality so that expert assistance can be obtained. To this end, some of the complications that can arise during foaling and shortly after birth are discussed in this section.

If a problem is suspected it is wise to contact the veterinary surgeon straightaway rather than 'wait and see for a while'. Mares do not take very long to produce their foals and the second stage can occur with very strong, intense contractions. If there is a problem at this stage, the situation can become life threatening for either mare or foal, or both, very quickly. Time is the critical element during an abnormal foaling.

Warning signs that something may be wrong may include one or more of the following:

- Appearance of the red placental membrane (allanto-chorion) at the vulva at the start of the second stage of labour instead of the white amnion.
- Appearance of the amnion without any sign of the foal's forelegs or head.
- Constant rolling, getting up and down, obvious distress from the mare.
- Mare failing to show any sign of straining, or does strain for long periods with no effect on the delivery of the foal.

In all but the first of these examples, experienced assistance should be sought straightaway as a matter of emergency.

In the first example, the immediate course of action is to rupture or tear open the red membrane that is visible with fingers or blunt-ended scissors and then get assistance. The red membrane that is visible is the allanto-chorion, which in normal circumstances will have thinned and ruptured at the area adjacent to the cervix, known as the placental pole, or cervical star, during the first stage of labour. The rupture of this membrane allows the release of the allantoic fluid, which signals the start of the second stage of labour and also acts as a lubricant for the foal in the birth canal. If the membrane is unusually thick, possibly because of disease or if it has released from the uterine wall prematurely, the pressure of the contractions may force it out through the birth canal until it is visible at the vulval lips.

The immediate problem in this case is that normally the foal cannot be delivered with the membrane intact and, because separation from the uterine wall has obviously occurred, the foal is being starved of oxygen until it is delivered and it can start to breathe for itself. Some foals are stimulated to breathe immediately the placental separation occurs meaning that they can drown in the amniotic fluids or they can suffocate or suffer brain damage through lack of oxygen if the membrane is not ruptured straightaway. It is not a common occurrence but one that requires immediate action. The biggest hurdle for inexperienced attendants to overcome is establishing that it is the placental membrane and not the uterus itself as many mistakenly think. The uterus cannot be pushed out in this way whilst the foal is still inside it, so the membrane should be ruptured immediately rather than waiting for a veterinary surgeon to arrive and do the same. This prompt action may save the foal's life; time wasted will almost certainly mean its death.

Many of the problems that occur during foaling relate to the position of the foal during birth (Figure 6.9). As we have seen, the foal has to turn from lying sideways or upside down during the last stages of gestation to extending the forelegs and head and coming into an upright 'diving' position. If the foal is unable to reposition itself correctly, perhaps because of a deformity, it will cause an obstruction in the birth canal, normally at the pelvic girdle and will not be able to be

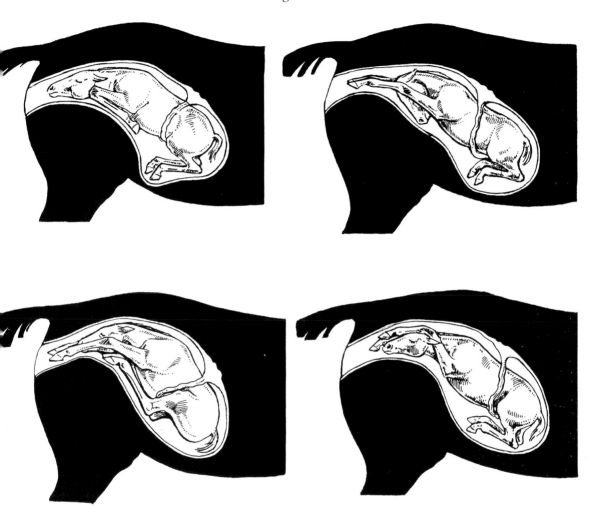

Figure 6.9 *Four examples* (reading from left to right) *of malpresentation of a foal during birth: foal presenting* (top, left) *with both forelegs back; foal presenting* (top, right) *with head back below the pelvic brim; foal presenting* (bottom, left) *with hind legs forward over the pelvic brim; and foal presenting* (bottom, right) *with forelegs over the head*

born without assistance. Malpresentation can cause severe damage to the foal or its death as well as cause damage to the mare's genital tract, so experienced assistance should be sought quickly. If the labour has been allowed to continue for a long time with the foal in the wrong position, it can be very hard to correct without damage to both parties. Very severe malpresentations may require the mare to undergo a Caesarean section to deliver the foal. This kind of surgery may not be available to all breeders, especially if the stud is in an isolated location, as it is a specialized procedure that requires full hospital facilities for the actual surgical delivery as well as for the aftercare of mare and foal. Caesarean section is not without complications, but the success rates are reported to be good. If a Caesarean is impossible, and the foal is dead, veterinary surgeons may remove the foal in sections through the mare's vagina. This specialized technique is termed a fetotomy.

Problems with the mare during foaling are also relatively rare but, again, prompt recognition and action from an attendant can make all the difference to the outcome. Conditions such as torsion, or twisting, of the uterus, rupture of the uterus or of the intestine can occur prior to or during labour. Torsion and/or rupture of the uterus or intestine will obviously cause extreme pain. With torsion, the mare shows signs of colic type pain and usually the veterinary surgeon will diagnose the problem quickly by rectal palpation. In some cases, the foal is unaffected and the mare is able to breed successfully in the future but, sadly, often this situation results in the loss of the foal and some degree of fertility compromise for the mare. Rupture of the uterus can be caused by an accident such as a fall, but often it is caused by a difficulty during labour that is not corrected quickly. A uterine rupture often causes the mare's contractions to stop and this may be the first indication that something is wrong. On examination, the veterinary surgeon will probably try to establish the location and size of the tear. If the tear has occurred during labour, the veterinary surgeon may decide to assist the prompt delivery of the foal manually, or to surgically deliver by Caesarean section. Rupture of the intestine may also occur during labour and affected mares show signs of shock with patchy sweating and increased heart rate. As with uterine rupture, intestinal rupture may also cause the process of labour to be stopped as the first main sign of something being wrong.

Other complications of birth may involve damage to the mare's genital tract by the foal during the labour process. Damage to the vagina and vulva is a relatively common problem in foaling mares. The damage can range from small tears that require no treatment to major lacerations that may permanently affect the mare's chances of breeding successfully again. One of the most significant lacerations that can occur is called a rectovaginal fistula (Figure 6.10). This occurs when the foal's feet go right through the roof of the vagina and penetrate the rectum of the mare during labour. If uncorrected, the force of the contractions can cause the foal's feet to tear the mare right along to the anus, making a common passageway between the vagina and the rectum. This situation often occurs because a mare has foaled unattended and could have been corrected with minimal damage if noticed early enough. Although a rectovaginal fistula can look horrific, it can usually be repaired by a veterinary surgeon although the mare may require further reconstructive surgery as the scarred genital conformation may predispose her to infection.

One of the most serious conditions that can affect mares during labour is that of haemorrhage. How serious it is will depend on the location of the haemorrhage and the blood vessels involved. Acute haemorrhage, where large amounts of blood are lost quickly, usually from a main uterine artery, are often fatal, whereas chronic haemorrhage, where small amounts of blood may be lost over a period of time, can be treated. Haemorrhage may be caused by direct damage during the actual birth, either by the foal or by a careless attendant. However, haemorrhage can also be caused by overstressing of the uterine arteries during gestation leaving them weak during the forceful process of birth. Exercise, age or stress straight after giving birth may precipitate haemorrhage in a susceptible mare. An important point to note is that a twitch should not be applied to the mare for 4–5 days after foaling as it may predispose her to uterine haemorrhaging by raising her blood pressure.

Haemorrhaging mares may display a range of signs. Initially a mare may be reluctant to rise after foaling, but then begin to roll and show signs of pain. At first these signs can be hard to distinguish from what is expected behaviour at the third stage of foaling, so as always, if in doubt seek veterinary advice. If the haemorrhaging is severe the

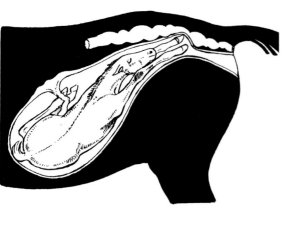

Figure 6.10
The sequence of events (reading from top to bottom) *that occur causing a complete rectovaginal fistula*

As the birth progresses (top, left), *one or both of the foal's forelegs can continue upwards rupturing the tissue between the mare's vagina and rectum, causing a fistula, literally a hole or tear* (middle, left). *If this position is not corrected immediately, the force of the mare's contractions will cause the foal's hoof or hooves to tear right through the whole length of the recto-vaginal tissue* (bottom, left) *until the hoof or hooves of the foal reach the mare's anus* (bottom, right)

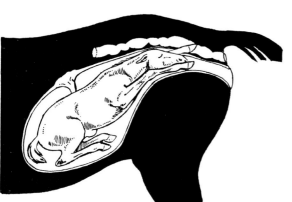

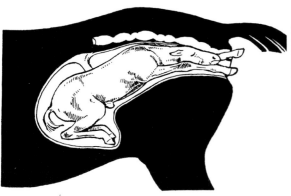

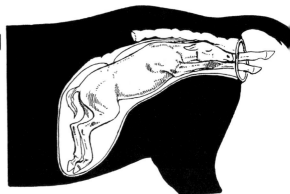

mare's mucous membranes will become pale, she will exhibit signs of shock, start to stagger about and usually death will occur before help can arrive. If the mare starts to stagger about in the stable the foal should be removed to another stable immediately for its own safety as the mare could collapse at any time. Extreme care should be taken, especially by inexperienced attendants, when trying to handle a mare in this situation.

Problems Affecting the Newborn Foal

During the first two days of its life, the foal will undergo several complicated physical changes as it adapts to life outside of the uterus; this is often called the adaptive period. It is during these first few days that many of the common newborn foal problems will become apparent.

Life outside the uterus is very different for the newborn foal. We have already discussed the way that the placenta provides nutrition, waste removal and protection for the growing foetus. Once disconnected from this support system, the foal must adapt quickly to managing all these roles for itself if it is to survive and thrive. These adaptations involve all of the foal's major systems – respiratory, circulatory, neurological, digestive and excretory – and some will occur within a few hours of birth, some take a little longer and that is why this period is critical. If the foal fails to make these changes for whatever reason its very life is threatened. From a breeder's point of view, it is important to be familiar with what is normal and to recognize when something is not quite right as soon as possible.

The normal heart rate of a foal is about 100 beats per minute when it is a day old, but at birth it may be 80 beats per minute rising rapidly to around 140 beats per minute as it puts all its effort into trying to stand. The respiratory rate of a foal is usually around 30–40 breaths per minute at rest within a few hours of birth. Body temperature is about 37.3–38.3 degrees Celsius and the normal foal is well equipped to maintain this with a layer of body fat and thick insulative coat.

Most mares will have healthy foals year after year and it is true to say that neonatal complications are uncommon, but breeders should have some familiarity with the most frequently reported abnormalities, and these are briefly discussed in what follows.

Non-infectious Causes

Meconium retention occurs when the first faeces become hard and impassable, causing the newborn foal considerable discomfort. The main signs are similar to those of colic and foals suffering from retention may often get into strange positions such as on their backs, as they attempt to relieve the pain. This condition is often not apparent until about 24 hours after birth and may occur even though some meconium has been passed. It is more common in colts. Giving foals enemas as routine shortly after birth can help with straightforward constipation, but a more serious impaction will require veterinary attention.

Rupture of the bladder is a rare condition in newborn foals but may occur following a traumatic birth. Again, the foal may appear normal for 24 hours and then shows signs of illness such as weakness and depression. Affected foals will also develop a swollen abdomen as the urine leaks from the rupture directly into the abdominal cavity itself. This condition can be fatal without prompt veterinary attention.

Another condition of the urinary system is patent urachus. The urachus is the tube that connects the foetal bladder to the allantoic membrane so that urine can be removed to form the allantoic fluid and obviously, this tube becomes useless after birth. The urachus is sealed when the umbilical cord breaks and the flow of urine is diverted to the urethra. In some instances, the urachus does not become sealed and a trickle of urine will continue to leak from the navel stump. Although not often life threatening, this condition requires prompt veterinary attention.

The respiratory system of the newborn foal is not commonly affected with non-infectious complications. However, damage can occur to the lungs following a traumatic foaling if the foal suffers from rib fractures. Given the foal's position during birth, it is obvious that the rib cage is very vulnerable to trauma, especially if the birth has not been normal. Lack of oxygen, asphyxia, in newborn foals may also occur through a traumatic birth. Even if the supply of oxygen to the foal has been affected for only a few minutes, permanent brain damage and/or death can occur.

The circulatory system of the foal has also operated with 'bypass tubes' during the time that the foal has spent in the uterus. The foetal heart has two bypass mechanisms to allow blood to flow to and from the foetus and placenta without having to use the non-functioning lungs. These diversions are also sealed shortly after birth to allow the normal flow of blood to and from the heart and lungs in the newborn foal. Occasionally, these diversions fail to seal after birth and the foal may show varying signs of weakness with blue tinged mucous membranes. As before, prompt veterinary attendance is necessary if this condition is suspected. Other defects in the heart structure are uncommon in horses.

Haemolytic disease is a serious circulatory condition that can occur in newborn foals. It is sparked off during pregnancy when a few of the blood cells of the foetus leak into the mare's bloodstream. The foetus is essentially a foreign body in the mare and it is only due to the 'override' mechanisms of pregnancy that it is not attacked by the mare's own immune system. In haemolytic disease, the contact between the blood cells of the foetus and mare provokes an immune response and the mare builds up antibodies to fight off what her body considers to be foreign. These antibodies are then transferred to the colostrum ready to help protect the foal against disease. However, this is one of those situations when the body reacts in the wrong way and the antibodies will be ingested by the foal when it sucks for the first time and immediately begin to attack the foal's own blood cells. This situation does not happen every time foetal and maternal blood cells come into contact during pregnancy, but only in rare situations where perhaps the genetic combination between mare and stallion predisposes it. Once a mare has produced a haemolytic foal, she will commonly do so again. It is possible to test the mare's blood and/or colostrum prior to foaling and if the results are positive the foal must be muzzled for 36 hours after birth; beyond this time the antibodies can no longer pass across the foal's intestine.

So long as the foal does not ingest the mare's colostrum it will remain unaffected, but it must be given colostrum from another mare, or a powdered alternative, until it can drink freely from its mother.

During the adaptive period, the foal's neurological system comes under an onslaught of stimuli. Normal foals cope well but those that have suffered some trauma during pregnancy or birth may fail to do so. These foals often display characteristic abnormal behaviour and, whilst they may have appeared normal at birth, within 24 hours they appear to lose some or all of the skills they had displayed earlier such as a suck reflex and the ability to stand.

Foals displaying these problems are termed as suffering from neonatal maladjustment syndrome (NMS). Sometimes, foals displaying similar signs are often termed as dysmature. In simple terms, dysmature foals basically fail to cope with life outside the uterus and may be weak, undernourished at birth and slow to produce normal behaviour right from the start. NMS is a term used more commonly with foals that appear normal at birth and then exhibit signs of brain damage within a few hours. The usual signs of NMS are convulsions, wandering, appearing to be blind, loss of suck reflex and coma. NMS is seen if the foal has been starved of oxygen for a period of time, either during pregnancy or during birth. The severity of the condition will depend on many factors such as the areas of the brain that have been damaged and the length of time that the foal was without sufficient oxygen. With careful intensive nursing foals suffering from NMS can make a complete recovery but the first few days are critical. Foals suffering from neurological disturbances, whatever the cause will require expert intensive nursing, as they will be unable to stand for themselves or suck from the mare. Some small breeders may not have the facilities to provide a sufficient level of care or find the expense of this form of intensive care financially prohibitive.

Physical Abnormalities

Physical deformities may be more common in some breeds, such as limb deformities in Thoroughbreds. Other deformities and limb weaknesses (Figures 6.11 and 6.12) may be inherited and mares that

Figure 6.11
A foal born with weakness of the limbs. Although dramatic to look at, many limb weaknesses are self correcting with careful expert management

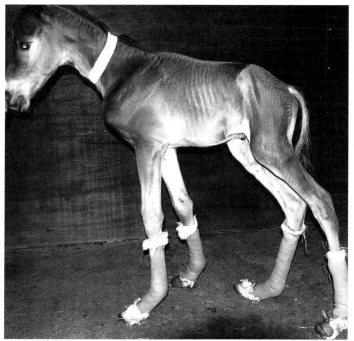

Figure 6.12
Applying soft bandages to the foal's legs is essential to reduce the risk of further pressure and friction injuries. The bandages also provide support

display deformities themselves and/or produce deformed foals should be removed from the breeding herd. Some deformities occur as an isolated case and may be due to the position of the developing foetus in the uterus. The infinite complexities of genetic combinations in any individual will mean that, from time to time, a fluke deformity may appear however careful the breeder has been with the selection of sire and dam.

Limb deviations and weaknesses are common in foals. Many look much worse than they actually are and with careful management, veterinary advice and expert farrier assistance many foals go on to develop completely normally.

Severe limb deformities may have caused problems during birth and the foal has a limited future if it is unable to stand or feed for itself. Foals with serious deformities of the limbs may never be able to develop properly and euthanasia is often the best course of action.

Aside from limb problems, some of the more commonly seen physical abnormalities may include cleft palate, where there is an opening from the mouth to the nasal passages due to the palates not forming correctly. Foals with a cleft palate often show signs of milk dribbling from the nostrils during or after sucking. Complications can arise from a cleft palate and, if suspected, the foal should always be assessed by a veterinary surgeon.

Entropion is a relatively minor condition that can occur in newborn foals where the eyelid, usually the lower lid but sometimes both, becomes inverted and the eyelashes constantly rub against the delicate surface of the eye causing scratching and irritation. If left untreated the surface of the eye can be damaged permanently leading to some degree of blindness. It is a condition that is relatively simple to treat but should always be assessed by a veterinary surgeon shortly after birth.

Umbilical hernias are often present in newborn foals, although they can appear later, when the foal is a few weeks old. The opening in the navel area that allowed the passage of the umbilical cord during development in the uterus normally closes during the last stage of gestation. If the opening does not close there is a chance that a loop of intestine can slip down through the hole – this is the hernia.

Small hernias can resolve themselves as the foal matures, although this may take several months. Larger hernias can cause complications such as strangulation, which occurs when the hole closes over the loop of intestine, cutting off the blood supply. Large hernias may require surgical treatment.

All hernias should be assessed by a veterinary surgeon so that advice can be sought as to the best course of action. Umbilical hernias are unsightly and this can be an important factor for breeders looking to sell or show their foals at an early age.

Infectious Conditions

In most normal circumstances, the antibodies passed to the foal from the mare's colostrum are sufficient to fend off challenges from environmental pathogens. However, as we have discussed, there are occasions when despite the foal receiving sufficient levels of good quality colostrum, there is a failure in the transfer system of this protection. This fact alone makes blood testing foals for IgG levels at about 24–48 hours from birth essential. It is a relatively inexpensive test and foals that indicate a low level of IgG concentration in the blood can receive a course of antibiotics as added protection. Some large studs give a three-day course of broad-spectrum antibiotics as a matter of routine for every newborn foal regardless of IgG level.

All infectious conditions are serious in newborn foals as their physical immaturity makes them unable to cope with all but the most simple of illnesses. In some cases infections may be present at birth, such as with EHV, and may also be ingested by the foal as early as the first few hours after birth, when the foal seeks the mare's udder but sucks at walls, etc. Infection can also gain easy entry via the stump of the umbilical cord. In the latter case, many studs will routinely apply antibiotic powder or spray to the stump as soon as the cord breaks.

Stable hygiene in the foaling area is the most important way of reducing infection risk to the newborn, as well as ensuring that the foal has received adequate protection from colostrum.

Infections in the newborn may cause similar signs to that of NMS with convulsions, weakness, loss of suck and coma. Any foal showing signs of disease at or shortly after birth should receive veterinary attention as a priority. Again, time is critical to the chances of a full recovery.

Newborn infections include septicaemias where the whole body becomes invaded by the bacteria or the toxins that the bacterium produces. Occasionally, foals may have become infected whilst still in the uterus, but commonly these types of infections develop shortly after birth due to the bacteria gaining entry through the mouth and navel of the foal. The infection may also gain entry through an injury to one of the foal's joints – setting up a form of septic arthritis. Any athletic future for an affected foal will obviously be in doubt.

Joint ill is a condition of foals that is commonly caused by a pathogen entering through the navel. Affected foals may have also suffered an umbilical abscess. Very young foals, a few days old can develop joint ill, or it can affect those three or four months old. Joint ill can be difficult to treat and often causes a high temperature, lameness and loss of appetite. In some circumstances, foals may suffer from permanent lameness as a result of joint ill. Veterinary attendance is vital to give the foal the best chance of recovery.

Diarrhoea is one of the most common conditions that many foals will suffer from at some time. It can range from a mild scour with no loss of appetite to a full blown debilitating diarrhoea with dehydration, weakness and collapse. Young foals because of their immaturity are very susceptible to dehydration and may become seriously ill in just a short period of time.

Long-term diarrhoea can cause scarring and ulceration of the gut and this may be permanent.

Foals commonly suffer from a relatively mild scour, often called a foal heat scour, about 7–15 days after birth. It generally lasts only a few days and affected foals may not appear ill. Infectious diarrhoea may be caused by pathogens such as rotavirus and salmonella and can cause varying degrees of illness in foals, with newborn foals being the most seriously affected. The signs, aside from copious diarrhoea, are weak and listless foals who are commonly off suck – that is, they do not feed at all from the mare.

Establishing the cause of the diarrhoea is important and this can be easily done by taking a faecal sample from affected foals for veterinary testing. All affected foals should be treated as infective until the test results are available, especially if a large number of foals are present on the stud.

Orphan Foals and Fostering

The sad loss of a mare during foaling or shortly afterwards can be a real blow to any breeding operation regardless of the size or type. However, a foal may also be termed as an orphan when its mother rejects it or its mother is physically not able to rear it. A newborn foal depends on its dam not only for immunity to infection through the colostrum, but also for the development of normal behaviour. Orphaned foals can suffer serious physical and psychological setbacks at an early and critical stage in their development; therefore, it is vital to plan a proper supervised care programme to reduce the additional stresses that the foal will have to cope with.

The immediate needs of an orphan foal will vary enormously depending on age and circumstances. The most important factor is immunity level. If the foal has been orphaned at birth, it will not yet have received any colostrum. Even if the mare has died during parturition, it still may be possible to use her colostrum and bottle-feed it to the foal. This method has produced very good levels of passive immunity in a high percentage of orphaned foals.

The orphaned foal may have suffered some level of compromise during a traumatic foaling and will therefore require special care in the immediate period following birth. Weak or depressed foals will normally require to be tube fed initially, or injected antibiotics given, as well as an enema to encourage normal evacuation of the bowel. Stronger foals will normally readily accept a bottle and can be put onto a regular feeding routine immediately – colostrum in the first feeds changing to a suitable mare's milk replacer as appropriate. As with any other foal during the immediate period after birth, careful monitoring is essential – perhaps more so with an orphaned foal who has already suffered additional stress.

Mares may reject a foal for a variety of reasons. Some mares having their first foal can become frightened and react aggressively to their offspring in their confusion; some mares repeatedly reject their foals year after year. As discussed in earlier chapters, it is essential to keep human intervention to a minimum during the initial period following foaling – particularly with a maiden foaling mare. The bond that is formed between mare and foal during this period is vital. A mare who has required attention following foaling, or perhaps her foal has required special care, may not form a strong bond and subsequently fail to maintain a sufficient interest and level of care towards her foal. There are many factors that can inhibit the normal response of the mare to her foal and great care should be taken to avoid a pattern of behaviour becoming established.

A very small percentage of mares may savage their foals – some causing very serious injuries. In these situations the foal should be removed for its own safety and the mare suitably restrained. Some mares seem to have a delayed acceptance of their foals, initially being aggressive and then becoming protective – this may be due to factors such as udder tenderness which is relieved once the foal has had an adequate chance to nurse. In the case of the aggressive mare, it is worth trying to get the mare to accept her foal by similar procedures to that used when getting a foster mare to accept an orphan as this is reported as successful in a good number of cases. If a mare is showing signs of increasing irritation and lack of interest in her foal all possible checks should be carried out to ascertain that there is not a simple reason to explain her behaviour. Mares can 'learn' to react aggressively towards their foals and, particularly with Thoroughbreds, this conditioning may result in her reacting in the same way at every birth resulting in her never being able to raise her own foals.

Orphaned foals need to be fed at least once every two hours, day and night during the first week. Hand rearing is extremely time consuming and, if there is no alternative, some handlers will try to get the foal drinking milk from a bucket or bowl very early on. Most foals will master this quite quickly, but it is a technique that should only be used if fostering has been decided against. There are many reasons why hand rearing may be the only option – the foal may be too weak or ill, a foster mare may just not be available or the owner is determined to hand rear the foal. The ideal method for raising an orphan foal is to find a suitable foster mare. Hand rearing is perfectly possible, but is not ideal as the foal will thrive better if fed and cared for by a mare; a hand-reared foal may become too humanized and, ultimately, difficult to train. Fostering of a foal can be time-consuming and immensely frustrating during the early stages, but is the best possible alternative to the foal's own dam.

Advertising for a foster mare can be done via the specialist press or, as used by many Thoroughbred studs, by contacting racing programmes on television for announcement during a televised race meeting. There are organisations such as the National Foaling Bank in Shropshire, who may be able to put the studfarm in touch with a mare owner who has lost a foal, as well as offer considerable advice and assistance. There are also people who specialize in providing foster mares for studfarms. This is usually on the basis of payment of a fee for the mare and the request that the mare is returned in foal when the fostered foal is weaned. These mares are normally of native or mixed breed and are kept specifically for fostering foals; some large studfarms keep one or two 'nanny' mares for fostering should the need arise. On occasion a studfarm may prefer to keep a valuable foal at home when the dam goes away to stud, particularly if this is overseas, to reduce the risks of injury or infection and the foal is then fostered on to one of the native mares.

Choosing a foster mare may not be easy, as it will very much depend on availability of a suitable mare at the right time. The most important factor is to try to find a mare that seems to have the right temperament for fostering. Some mares that have lost their own foals will not accept another as a replacement and can be extremely aggressive towards the new foal. The breed of the mare is immaterial if her temperament and attitude is suitable, but normally native or heavier breeds tend to adapt best. Size of the mare may also need to be a consideration – fostering a large breed of foal onto a pony mare may be fine for the first few months, but we all know how quickly foals can grow. Feeding from a very small mare may become difficult for the orphan within a few weeks and the foal may need to be weaned considerably earlier than planned.

The other vital consideration in looking for a foster mare is to establish her health status and as to why her own foal was lost. Any risk of disease needs to be treated very seriously to avoid the risk of bringing infection onto the stud. Normally, as it may take up to 7–10 days to get full laboratory results back, a foster mare is kept in isolation with the orphan foal until confirmation of her status is available. If isolation facilities are not available, it is worth arranging for the orphan foal to be moved to a suitable place with the mare, away from the studfarm until results are known. As most foals are born during the breeding season, it is commercial suicide to allow such a mare contact on the main stud until full results are available. Obviously, not all foals die due to infection and if it can be established that the foal died or was euthansed due to a non-infectious cause such as physical deformity, then obviously the risk is reduced. However, it is important to allow time for the normal tests to be carried out for infections such as EVA and CEM. All horses coming onto the studfarm should be tested for EVA and all mares, regardless of whether they are to be bred or not, should be tested for CEM. In such situations, the advice of the stud's veterinary surgeon should be sought.

The old practices of putting the dead foal's skin on the orphan foal or covering the orphan in the mare's birth fluids or urine are thankfully no longer commonly used. Not only are such practices unnecessary, they exacerbate the stress the orphan foal is already suffering as well as increasing the risk of secondary infections.

Once the mare is available for the foal, the next most important factor is maintaining the safety of the foal at all times and providing a suitable quiet environment to allow fostering to take place. In the ideal situation, using a fostering pen will allow the foal safe access whilst the mare is restrained in a small stall and can safely get used to the foal's presence. Some foster mares need little or no introduction to the orphan foal as their maternal instinct is very strong – careful observation will give the handlers some idea as to how the mare is likely to react. It is vital not to take any chances at this early stage and to be certain that the mare is never given the opportunity to harm the foal in anyway. Successful fostering needs a handler who is able to take the fine line between not allowing enough contact between the mare and foal and allowing too much and putting the foal at risk. It should be understood that successful fostering usually needs a good team of experienced staff. If a fostering pen is not available then hay or straw bales can be used to make a makeshift barrier – but as before, safety of the foal must be given priority.

The orphan's age will also affect the way that fostering can be set up. A very young foal may be slow to latch on to the mare, especially if bottle feeds have been necessary whilst a foster mare was found. A young foal may also be at more risk of injury by an impatient mare, being clumsy in attempts to obtain milk direct from the udder for example. This is a time when the patience of the handler is vital. Forcing the foal to suck will only confuse matters further, as will using sweet substances on the mare's udder as an encouragement. Instinctively, the foal learns to suck properly, but it can take some time to establish and fostering is very labour intensive at this stage. If the foal is becoming very weak and has yet to suck, it may be necessary to give a supplement feed, but it is worth bearing in mind that this may work against the fostering if the foal is then not hungry enough to want to try.

A young foal should be encouraged to the mare very regularly during the first few days and nights and encouraged to suck. The mare should be adequately restrained at this time, but if possible allowed to see the foal and, if she is quiet, to touch the foal. Allowing the mare contact with the foal will encourage bonding to take place and reduce the levels of stress on the mare; however, she must not be allowed free access to the foal at this stage unless she can be completely controlled should she react aggressively. Mares can cause horrific injuries to foals and should be closely monitored at all times during the early stages of fostering.

The whole fostering procedure can take up to a week to allow the mare time to fully accept her new foal. If the mare has shown only interest in the foal and is happy to allow the foal to suck from her unrestrained, then relatively free access to each other in the stable should be allowed. It is wise to continue to have some form of control over the mare until it is certain that she will not attempt to reject the foal – once she displays protective behaviour, the adoption can be considered successful and it is unlikely that she will reject the foal from now on.

Once this stage has been achieved then mare and foal should be left alone to further the bonding process. Closed-circuit cameras can be invaluable in a situation like this as too much human interference may affect the success of fostering. The mare and foal can be safely monitored with no knowledge that they are being watched and their behaviour and reactions will be more natural. It is essential that they are still observed as much as possible; there have been occasions when a mare will reject the foal as soon as she thinks she is left alone but be happy with it when people are present.

As soon as the mare and foal are settled together, it is useful to try putting them out together in a small nursery paddock. Even if the mare and foal are cleared from any need for isolation, it is important that they are not put out with other mares and foals at this stage – foster mares can sometimes be less protective of their foster foals and there is a danger that the foal may wander to another mare and be hurt.

The normal behaviour of the mare towards the foal – calling, allowing it to suck and keeping it close to her, etc., can be monitored in a safe paddock. Monitoring allows the handlers to get some idea of how the mare will be with the foal when she is ready to be turned out with others.

There are no set time limits for fostering – as we have seen, there are too many variations. Some mares will accept an orphan foal completely and immediately, making all of the precautions and procedures detailed at the beginning of this section unnecessary. Others may take three or four days before they will allow the foal to suck unrestrained. Some mares are determined not to accept the orphan even with the best efforts of the handler. These mares cannot be left alone with the foal as they attempt to force the foal away even when a handler is present. It is worth trying for two to three days if the foal can suck from the mare, but if she still shows little to no interest then it is probably best to call it a day. Sadly, not all fostering is successful – however, if another mare is available it is worth trying again.

The most important factor is that a suitable mare is found as soon as possible after the loss of the foal's own dam. The longer the foal is hand reared with bottle feeds the more difficult it can be to get it interested in sucking from a mare. Sucking is instinctive behaviour, but if the foal has imprinted onto a human with a bottle it can be hard work to change its mind. Sucking from an udder is also more difficult for the foal than sucking from a bottle; the foal may be reluctant to put the effort into trying, causing irritation in the mare.

Always reduce handling of the foal to a minimum during the waiting period for a foster mare to become available; doing this will help to reduce the strength of bond that will form between the foal and human attendants. Some studs will only tube feed newborn orphan foals to avoid any further confusion, although obviously this is normally only a short-term option. It is worth noting that foals born prematurely may require tube feeding as they usually display an insufficient suck reflex to feed from a bottle.

Hand rearing will also mean that the foal will have to rely on milk formulas rather than mare's milk. There are some very good milk replacers on the market, but none can match the real thing. Some hand-reared foals develop a level of digestive upset when fed milk replacer and such individuals need extra care. Normally, hand reared foals are introduced to milk pellets and creep feeds as soon as possible, but as with all diet changes this will need to be done gradually and such feeds are not always suitable for the very young foal.

Foals will generally not consume very much grass or hay in the first few months of life. It is important that any food they are offered is easy to digest as their digestive systems are still developing in every sense of the word. Setting up a creep feeding system in the foal's paddock or stable will enable the gradual change from milk to concentrates.

Regardless of the circumstances that resulted in the foal being orphaned, careful consideration should be given to its psychological well-being. The foal may have been bred to hopefully be at the top of a chosen sphere – be it as an elite athlete racing or eventing, or as a show horse. Therefore, it is vital that every care is taken to ensure that the fact that the foal begins life as an orphan is not allowed to compromise its potential long term. Orphaned foals are more prone to infections and extra consideration should be given to their early management.

The best lesson any handler or manager can give to an orphaned foal is to encourage it to be a horse. If hand reared, the foal may have little or no idea of the complex social behaviour expected when in the company of other horses. Some displays, such as biting, kicking and rearing, are natural forms of horse behaviour and as orphans tend to treat humans as their peers, they can become dangerous if not disciplined from an early stage as to what is acceptable behaviour and what is not. Even if it has not been possible to foster the foal onto a mare, it is vital that every effort is made for the foal to have a suitable horse companion – ideally of a similar age – so that learning to develop normal equine social behaviours and attitudes begins as early as possible.

7 The Foal from Birth to Weaning

The first year of the foal's life is an important one in terms of growth and development. For many breeders, the aim is to produce young-stock for early sale as foals or yearlings and it is during these early months that careful management can make all the difference to the foal's physical and mental development. The subject of foal management is complex and it is impossible to cover all aspects in one short section. Here, we discuss only the basic principles of early foal training and care.

Early Handling

The early months of the foal's life are an ideal time to start the basic training of the foal – such as wearing a headcollar and being led, being groomed and receiving farrier attention.

Many studs will fit leather foal headcollars as soon as a few hours after birth. On studs with a large number of foals, the headcollars are also fitted with identification tags so there is no risk of mismatching a mare and foal. The headcollar should always be properly adjusted so that the noseband is loose but not too low down on the nose and so that the headpiece cannot slip down the foal's neck. Leather headcollars are always preferable to nylon, particularly for foals, as leather breaks more easily should the foal become caught up. Headcollars should be thoroughly cleaned and checked prior to use each year, as some mares may become distressed if the headcollar smells of another horse.

Teaching foals to be handled and led should be done on a daily basis. Foals tend to react instinctively to new experiences and their first thought is often to run away – this is quite normal behaviour as the flight reaction has saved the horse from predators many times in the past. However, in the domestic environment this sort of reaction has to be overridden by careful training and lots of patience. Some individuals are naturally more nervous than others and will require extra time and care to achieve the easiest levels of training. Starting to teach foals still with their mothers makes everything much easier, regardless of the foal's individual temperament, as the foal is programmed to learn from its mother, picking up on the almost imperceptible cues that she gives.

Young foals are not led directly from their headcollars initially. They are usually 'cradled' by the handler, who puts one arm around the foal's chest and another around its quarters. This allows the handler to have a good level of control over the foal, which can be important when leading it to and from the paddock, etc., where new experiences/noises could startle it. Foals naturally want to follow their mothers and they can usually be easily led, either to one side or just behind the mare. Once the foal has learned to be cradle-led alongside its mother, the handler can start to teach the foal to lead from the headcollar. This is done by the handler leading the mare on her off or right side and leading the foal on its left or near side. The handler will normally not use a lead rope at this early stage and will instead just hold the foal's headcollar gently with two fingers through the back of the noseband. There should also be another handler to walk to the rear of the foal and who can help to encourage the foal to walk up with its mother. To avoid the risk of a handler being hurt by a kick, care should always be taken to not stand directly behind the mare or foal during leading. Standing to one side also means that both horses can clearly see the handlers.

In the wild state, young foals would always follow on behind their mothers and not go out in front – the mare is, of course, the foal's main protection. Training the foal to walk up alongside its mother may take a little more encouragement with a nervous individual, but it is an important mental hurdle for the foal to overcome. The domesticated horse has had to learn to override its fear of new situations and to not run away first and ask questions later. Trusting the judgement of the handler is a stage that all horses need to reach to allow training to be successful.

The young foal is very impressionable and learns a lot from its mother and, as it gets older, from other foals it has contact with, as well as from its human handlers. A bad experience during early training, perhaps due to an overly forceful handler, can adversely affect the foal's whole attitude to future training. The foal for its small size is very strong and physical battles should be avoided at all costs. Force never has a place in training horses of any age and careless handling can very easily lead to foals becoming injured physically and emotionally.

Catching young foals in the field requires, initially, at least two handlers. The mare should be led up to the fence line and the foal encouraged to stand between the mare and the fence, a position it will naturally put itself in. The handler should talk to the foal all the time to reassure it. Once the foal is calm, the handler leading the mare should move the mare closer to the foal and the fence line. The free handler can then approach the mare and foal, encouraging the foal to move up to its mother's head, the mare's handler can then take hold of the foal's headcollar and lead them forward calmly. Great care should be taken the first few times the foal is caught in this way as foals can often panic if they feel cornered and may try to jump through the fence. A fully boarded fence is the ideal, even if only a section is boarded, or alternatively a close barred gate may be suitable. During catching up the foal should be given as much time as possible to go up beside the mare and not be rushed or crowded. Some foals go through a stage of being difficult to catch and, once learned, this behaviour can waste a lot of staff time as well as increase the risk of the foal injuring itself. If there are a number of other mares and foals in the field, it is wise to remove them leaving just the difficult foal and its mother. Sometimes just the others leaving is enough to change the foal's mind, but if not the handlers can now concentrate on the single pair rather than having to keep an eye on the other mares and foals at the same time. Handler safety is always important and all staff should bear in mind that mares with foals at foot can be very protective. Even with a group of familiar mares and foals, handlers should avoid standing between the mare and her foal as this may be threatening to the mare.

Aside from leading, foals benefit greatly from being groomed gently with a soft brush. Foals adore being scratched, especially at the top of their tails or at the base of their manes as it mimics the mutual grooming behaviour that is a common part of the friendship bonding process with horses. Gentle scratching can help to calm an upset foal and is a good way of getting nervous foals used to being touched.

Teaching foals to have their feet picked up is another essential part of early training. Foals initially find it hard to stand on three legs, so each foot should be lifted only slightly off the ground for a very short time on the first few occasions. If the foal becomes used to having its feet picked up every day, perhaps to have them routinely picked out after coming in from the paddock, this will greatly assist the stud farrier.

Exercise and Behaviour

Foals develop and benefit both physically and mentally from exercise. Free exercise within a group of other mares and foals, also allows the youngsters to learn the social skills they will require for later in life when they are weaned from their mothers.

When they are very young, foals will start to develop play relationships only with their mothers and will not voluntarily stray more than a few steps from her side. By the time the foal is about two or three weeks old, it may start to gallop in circles around the mare, getting a little further away each time as it gets braver and it is at this age that foals often display mutual grooming behaviour with their mothers. In the early days, foals will rarely play with each other and those that do venture too close to another foal are usually chased away quite violently by the mare. Overly bold foals can be hurt by a protective mare at this age as they may not move away quickly enough and do not, as yet, understand the body language the mare gives to stay away. Foal injuries such as fractured jaws incurred by kicks can be relatively common in this age group, especially if a mixed age group of mares and foals is turned out together.

By the time the foal is about four to six weeks old, it will start to interact a lot more with other foals, so that by the time it is about eight weeks old it may spend as much as 50 per cent of its playtime with other foals rather than with its mother. Those foals with no playmates of a similar age will continue to play with their mothers for much longer than normal.

Young foals, like many other babies, spend a lot of time resting and sleeping. Even when out at grass, they will play for a while, rest/sleep and feed in short bursts. A foal lying down for a long period may indicate that all is not well, lying upside down or in any other unusual position is also an indication that the foal may be ill. Healthy foals generally sleep laid out flat on one side or resting up on their chests – often they will stay lying down even when approached so this is not generally a cause for concern unless there are other signs such as rapid breathing, sweating, etc.

Nutrition

Young foals will normally suck from the mare every 30 minutes or so. Very young foals may suck more often, possibly sucking for several minutes each time (Figure 7.1). If disturbed, foals will almost always get up and go straight to the mare and feed – a foal that is reluctant to feed may be unwell.

As the foals get older, they will start to suck less frequently from the mare, until ultimately they are weaned from her altogether. (Weaning is discussed in detail later in this section.) In the wild, the foal will normally be weaned a few weeks or days before the mare is due to give birth again – usually when it is about one year old. In the domesticated environment in most cases, the foal will have had access to

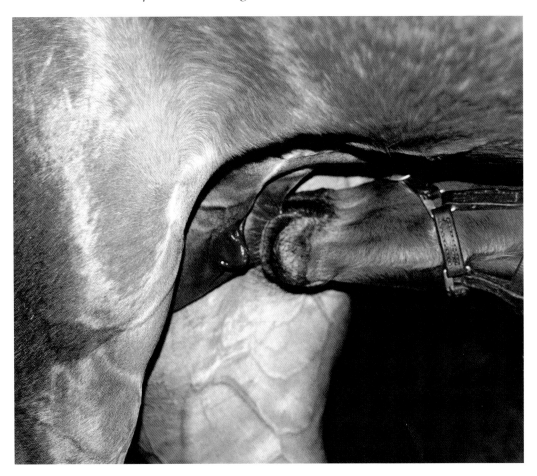

Figure 7.1 *Young foals will suck frequently, often for several minutes at a time*

concentrate feed and hay alongside its mother for many months and will have become nutritionally independent from her much earlier. Such foals may still continue to suck from the mare at about six months of age, but often this is due more to an emotional need rather than a physical one.

Mare's milk provides all that the foal needs in the first few weeks of life. The peak of lactation occurs generally at about 6–8 weeks after the foal is born and this is usually when the mare's nutritional needs are also at their highest to keep up with demand. Many mares, dependent on factors such as breed, time of year, quality of grazing, etc., can provide for their foals without the need of additional concentrate feeds. However, on Thoroughbred studs, where foals are often born early in the year before the spring flush of grass, some form of concentrate feed is essential.

Foals learn all sorts of behaviour from their mothers and most of this learning is done by copying. The young foal is also enormously curious and will investigate everything in its environment. If the mare is fed concentrates in her stable, the foal will often stand up next to her and try to eat with her, and will also copy her grazing behaviour in the paddock. As the foal gets older, it will start to eat an increasing quantity of concentrate feed with the mare as well as grazing during the day. Some studs like to provide creep feed for their foals from a very early stage. Creep feed is a special concentrate for foals, usually in the form of small pellets that are easy to eat and can be provided in specially designed feeders in the stable and paddock so that only the foals can get access to it. Most foals will not require additional concentrate feeds until they are much older, but if the mare is not providing enough for the foal and the foal is in poor condition, it should receive supplementary feed to avoid the risk of developmental problems later on.

Teeth

Horse teeth erupt as a gradual process that occurs at almost the same time in all individuals. This enables a horse's age to be established by looking at its teeth. Foals, like many other babies, develop milk teeth first, which will be gradually lost to make room for the permanent teeth as the foal grows and matures.

Foals are generally born without teeth, but they will cut the central incisors at about one week of age. The middle incisors appear later at about two months of age, with the corner incisors appearing later still, at about eight months of age. The baby pre-molars, or back teeth, appear at about three weeks of age. Wolf teeth, small pointed teeth that appear in front of the pre-molars, erupt at about eight months. These wolf teeth are not seen in every individual and are rarely seen in fillies. In Thoroughbreds, wolf teeth are often removed as they serve no purpose and can cause pain when the horse starts to wear a bit.

Routine Care

Farrier

The foal should receive regular farrier attention from as early as two weeks of age, especially in the fast growing breeds. Attention to the feet is one of the most important aspects of routine care and, sadly, is an area that is often overlooked. Expert farrier attention allows a breeder to evaluate each foal's structural development and allows for early correction of minor problems before they become major ones. Foals are growing so quickly during the early months that limb problems can become serious very quickly if left unchecked. The growth plates at the end of the long bones of the limbs are active only for a set period of time; once the growth plate has closed any deviation cannot be corrected by remedial farriery alone.

Foals that have limb deviations or minor conformational faults should receive very regular farrier attention throughout the whole of the first year of life. With an expert farrier, even severe limb deviations can be corrected with specialist remedial trimming. The bones in the foal's limbs are constantly growing throughout this time and quite dramatic corrections can be made successfully with a high level of care. Remedial work is not practiced by all farriers and it may be necessary to seek veterinary recommendation to find a farrier who specializes in foals and youngstock. Many minor limb deviations in foals are self-correcting and veterinary advice should always be sought before a course of remedial trimming is undertaken.

Vaccination

We discussed earlier the need for any stud to have a strict, carefully planned vaccination programme for all their horses from foals to veterans. Foals will have some early protection from infection from the colostrum they received from their mothers at birth, but many studs give their foals additional early vaccinations against some conditions such as tetanus.

Foals are more susceptible to disease than adults and care should be taken to reduce the risk of unnecessary exposure. This means keeping them away from other horses of varying age groups, except of course from their mothers. Weak or undernourished foals may be even more susceptible to the challenge of disease and foals in these categories require even higher levels of care.

The foal will gradually develop is own protection over time as its immune system matures and becomes capable of producing antibodies. This is often not complete until the foal is several months of age.

Routine programmes for foals are usually similar to those for adults, with influenza and tetanus being the main vaccinations given. As the foal is starting a vaccination programme for the first time, the injections need to be given at the proper intervals to produce the correct level of antibody protection. Some studs that keep a closed herd of horses, in that the horses do not come and go on and off the property, do not vaccinate their foals in the first year in the belief that the foal's immune system is not sufficiently mature to make antibodies and vaccination is, therefore, pointless. These studs start their vaccination programme once the foals have become yearlings. This theory is not standard practice on the majority of studs and those who compete with their foals or sell at foal sales would be required to produce vaccination documents as an entry requirement.

Parasite Control

Worming foals is another vital aspect of routine care. Foals are susceptible to certain types of intestinal parasite, such as *Strongyloides westeri* which can be transferred to the newborn foal through the mother's milk (adult horses have a level of immunity to this parasite).

Foals are vulnerable to heavy infestations of parasites and, if left untreated, their digestive systems can become permanently scarred and ulcerated.

Regular parasite control is important for foals and should begin when the foal is one month of age. Some of the modern wormer drugs are effective for up to 12 weeks and work by being stored in the fatty tissue of the horse. These wormers are unsuitable for foals under three months of age as they have insufficient body fat. Again, veterinary advice should always be sought if there is any doubt over a parasite control programme for foals.

Along with regular worming, paddock and stabling used for foals and weanlings should be subject to high standards of management to further reduce the risk of infection.

Weaning

Weaning from the mother is a stage that all mammals go through – some soon after birth and others several months or even years after their birth. Equines normally wean their offspring naturally after about a year, or shortly before the birth of the next foal. The bond between a mare and her offspring can remain very strong for many years, especially if they are kept in a small close-contact group, but the mare will rarely permit the foal to suck once the weaning transition is made.

For commercial studfarms weaning is a necessary procedure performed each year to ensure the optimum condition of both mare and foal. Most mares are bred each year and are normally four or five months pregnant with the next foal by the time they are weaned. Weaning at this time gives the mare a chance to regain any condition that she may have lost by nursing and allow her to concentrate on providing for the pregnancy. Additionally, the quality of the milk she is producing will be decreasing as the foal should be already receiving the majority of its nutritional needs from supplemental concentrate feeds and/or grazing.

With all that said, weaning is nearly always a very stressful time for both the mare and the foal. There are several weaning methods and each has its advantages and disadvantages. The main considerations for choosing a suitable method for an individual farm will be facilities, staff, number of mares, age of the foals, and condition of both mares and foals.

When the foal is weaned its life changes immediately and dramatically. It has lost not only its mother but the physical and psychological support a mother provided. Preparation for weaning is therefore vital. There are occasions where a mare may be lost suddenly, or the foal becomes ill and requires to be weaned to be cared for properly – it is impossible to plan for every eventuality but in most normal circumstances preparation for weaning cannot ever be considered unnecessary.

Foals should be given access to some form of supplemental or creep feed several weeks before weaning is planned. They will have already been showing an interest in any concentrate feed that their mothers may have had and will have been making attempts to eat with her from an early age. Creep feed can be provided just for the foals in a variety of ways. The individual hand feeding of each foal would be prohibitively time-consuming and labour intensive so other methods are much more common. These include providing a separate feed bowl in the stable that both mare and foal are housed in; the bowl should have narrow bars to permit only the foal to access the food.

For paddock creep feeding – perhaps the ideal way of providing feed for foals – the use of a creep feeding pen is common. The pen is made of either timber or metal and is constructed so as to only allow the foals access through narrow and/or low entrances (Figure 7.2). The single most important factor with any structure of this type is that it should be safe. The pen should be large enough to allow all the foals access and space to feed. The feed can be provided in either feed bowls or troughs within the pen – far enough in to stop any mares from being able to reach it. And it goes without saying that the structure should be sturdy enough to withstand the efforts of any mare who wishes to try and reach the food; it is surprising how determined some greedy mares can be.

Normally, the pens are set into the ground to provide the necessary strength for the structure, or existing fence lines can be utilized if

Figure 7.2 *A creep feeding pen*

practical and safe. Even when the foals are only a few months of age they already have a social structure and ranking system so care should be taken to ensure that there is sufficient space and bowls for each foal to eat without being bullied by another.

Creep feed is normally provided on an ad-lib basis as the foals will rarely over eat. However, each group of foals should be observed carefully during the first stage of introducing a creep system to ensure that each foal is taking full advantage. It is rare that foals will have to be shown the food inside the pen as they are naturally so curious, each stage of the pen construction will have been monitored closely!

The foals will normally start to eat well within a week or so of the creep feed being provided and with this increase in their concentrate diet, they will naturally begin to spend less time sucking.

Foals can be weaned completely from as early as three months of age and some farms report that these foals grow more quickly than those weaned at six months. However, the emotional stress on such a young and immature foal is greater than that suffered by an older more independent individual. Research suggests that by the time foals are yearlings the growth rate difference has reduced sufficiently as to be negligible, so early weaning may not be as productive for routine use as was once thought.

Large studfarms normally establish groups of mares and foals according to foal age, which assists with reducing some of the stress of separation for both groups. The foals will already be familiar with each other and there should be less risk of bullying as tends to happen with a group of mixed age weanlings. The mares will also be established as a herd group and normally will settle more quickly than if put together with an unfamiliar group. However, on smaller farms this may just not be practical and all the foals may need to weaned together regardless of age. If this is the method to be used, the weanlings should be carefully monitored for a long period to ensure that all are eating well and not being intimidated by older foals.

Traditionally, mares and foals were separated completely with the mares being moved to a far part of the farm and the foals generally left alone in a stable for a few days – some farms still prefer to keep weaned foals in a darkened stable by closing the top door and windows. This is about the most traumatic way of separating the pair. It takes anything from three to six days of complete separation for the mare and foal to settle and it can be an anxious time for the handlers too as the mares and foals can become very distressed and may injure themselves. Foals may also cease to eat during this period and can lose considerable condition, making them more susceptible to infection and, in some cases, causing considerable setbacks in their development to date.

If this method of separation is to be used it is best to separate the pair at the end of the day when the foal has had the benefit of paddock exercise all day. A tired foal is less likely to injure itself than one that has been confined. Bearing this in mind, foals that have been

confined following weaning should be released in a safe, small
enclosed environment when turned out first to allow them to 'let off
steam' before being moved to a larger turnout area. For these reasons
other methods are now more commonly used.

Familiarity will help to reduce the stress to the foal – maintaining
some of the elements of the foal's normal environment will also help.
For this reason, gradual weaning is currently the most common
method of separation. This process takes much longer than the traditional methods but is much closer to the ideal for all concerned. One
or two mares will be removed from an established group at a time.
Their foals may call and show some levels of anxiety for a short while,
but generally they settle down quickly when they are back with the
remaining mares and foals. The fact that the other horses in the group
are all familiar and are not distressed greatly helps the weaned foals
to return to normal routines. The mares are generally taken to another
part of the farm and may also show some levels of anxiety and distress for a short while. This is generally short-lived and some mares
seem almost relieved to no longer have the responsibility of their
foals. Additional mares are removed at intervals of a few days to a
week until the group is completely separated. Some breeders believe
that this method only prolongs the anxiety but research has shown
that this is not the case when compared to traditional methods.

Another method used, more commonly on smaller farms, is that of
gradual separation where the foals are put into stables or paddocks
close to the mares. This enables the contact between both groups to be
maintained but prohibits sucking. The foals are reassured by the close
presence of their dams and this method is extremely useful if there is
no way that the two can be kept out of sight and sound of each other.
However, the fencing does need to be particularly well maintained to
ensure that it is of suitable type and strength for this purpose.

Mares generally settle more quickly than foals but this will vary
with individuals and also be dependent on the number of foals she
has had before. The most important aspect of weaning for the mare is
encouraging her milk supply to cease. Care of her udder is therefore
paramount to avoid the risks of mastitis and secondary infections.
Decreasing the nutritional quality of her diet for the initial period is
helpful – some farms use relatively sparse and bare paddocks for the

weaned mares. Plenty of free exercise will also help to reduce the size and discomfort of a full udder – hand milking should be kept to an absolute minimum as the more the mare is milked the more milk she will produce – the pressure of a full udder will actually decrease and eventually stop milk production completely. Once the mare has completely 'dried up' then the quality of her feeds and/or grazing is usually increased.

8 | Stud Design and Administration

In an ideal world, the perfect stud design combines the right balance between practicality and aesthetics. Thoroughbred studs are perhaps the most palatial and the design ethos combines functionality with some of the most beautiful layouts seen in the horse world. However, on many of the large Thoroughbred studs, expense is not a prohibiting factor as it can be to the majority of horse breeders.

Whatever the bank balance, all horse studs will have common requirements, which apply to those dealing with small numbers of horses at the lower end of the market and those catering for hundreds at the top end. In this chapter, we discuss the basic principles behind stud design and selection.

Layout and Facilities

In the UK, many studs are set up by expanding or adapting existing properties, often from those originally designed solely for agricultural purposes. One of the most important factors when choosing a potential stud property is the land. Land is the foundation of the stud and should always be a main consideration. An area such as Newmarket in Suffolk has become a heavily populated horse area in part because of the land type. The predominantly limestone soil is well draining and fertile – ideal for the fast maturing Thoroughbred.

Studs must have turnout for their horses and that turnout must be suitable for use all year round. Horse properties set up for other purposes may be able to get by without turnout facilities, but for a stud it is essential.

Other factors to consider when assessing a property are good water supplies, drainage, supply of utilities such as electricity, ease of access, planning restrictions in the area and availability of extra space to expand in the future. The type and size of existing buildings should also be assessed, as adaptation may not be the most cost effective way of starting the stud. In many cases, unless the property has already been used for horses, many of the buildings may need to be removed and suitable structures built. Older properties tend to have low roofs in what was originally cattle housing and the partitioning may be too small for use with horses. The more modern farm buildings such as American style barns, however, are commonly adapted for horse use and with expert advice can provide excellent facilities for a stud.

On large studs, access to the paddocks should be as close to the stabling as possible to reduce time in moving the horses about and also to provide quick emergency turnout in case of a fire. The use of walkways between paddocks is common on Thoroughbred studs and they serve many functions. Between paddocks they reduce the risk of horses being able to reach one another over the fencing, an important consideration on commercial studs where a large number of visiting mares may be dealt with during the short breeding season. Walkways also reduce the chances of a horse getting loose when being led to and from the paddock, a factor that is important on any horse property, but particularly on studs where youngstock and stallions can be highly strung and excitable. If the walkway is suitable, the horses can also be moved from one paddock to the next by being driven, rather than having to be caught up individually.

Planning the layout for buildings should be carried out carefully. Correct planning and design will reduce the overall running costs of the stud by reducing the need for unnecessary labour and expense in the day-to-day running. It can be beneficial for the breeder to visit other established studs of a similar type to get some ideas and advice, to see what works and what does not before committing to any set plan. Any plan for a new stud will be drawn up taking into account the aim of the operation. If it is to be a small stud with no stallions, it will need only stabling for the mares and youngstock as well as foaling facilities, a tack room and storage areas for feed and forage. A public stud, standing one or more stallions, will require additional stabling for visiting mares, stallion boxes, teasing and breeding facilities, veterinary examination areas, isolation facilities and a stud office. Specialized businesses such as those providing AI and embryo transfer services will probably require additional buildings, such as laboratories.

Stabling

Stabling is one of the most significant long-term investments on any stud. Proper construction methods should be combined with a strong foundation, high-quality materials, proper insulation and ventilation. In almost all areas of the UK, the type of structure that can be built or adapted for horse use is subject to planning regulations and the local planning authority should always be approached well before any building work starts.

The stabling should be planned with consideration to ease of use as well as the health and safety of both horses and staff (Figure 8.1). Modern stud stabling is often in the form of barns, with, on larger properties, a separate barn being set up for each horse group: that is, yearling barn, barren mare barn, etc. Barns can also work well for mares and foals, but are not commonly used for stallions, who tend to be more settled with traditional stabling.

The design of a barn will depend on the type of horse to be housed and the financial and space restrictions. Many barns have feed/forage stores attached, as well as muck pits, but due care should be given to reducing any risk of creating a fire hazard or compromising stable hygiene. The main disadvantage with the standard barn system is that all the horses share the same air space, which increases the risk of infection transmission from horse to horse.

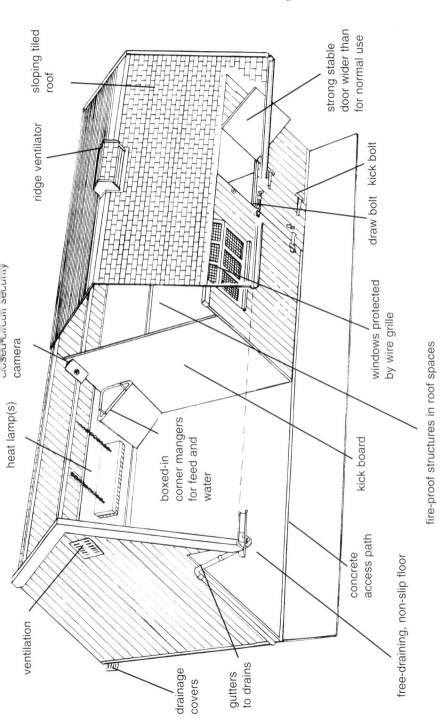

sloping tiled roof

ridge ventilator

closed-circuit security camera

heat lamp(s)

ventilation

drainage covers

gutters to drains

free-draining, non-slip floor

concrete access path

boxed-in corner mangers for feed and water

kick board

fire-proof structures in roof spaces

windows protected by wire grille

draw bolt   kick bolt

strong stable door wider than for normal use

Figure 8.1 *A typical stable used for broodmares*

The size of the stable will obviously depend on the type of horse being bred and the length of time that the horse needs to be stabled. A pony stud, for example, may only need stables for the ponies to stand in for a short time as they live out all year round, as compared to a Thoroughbred stud where the horses tend to be stabled at night and often part of the day as well almost all year round. The usual size for a stable is about 3.65 m x 3.65 m with a height of 3.05m from floor to rafters. Stables for mares with foals at foot are generally over 4 m x 4 m and those for stallions larger still.

The wall construction of the stables is often of concrete block, mainly faced with timber on the inside wall. The inner walls should always be smooth and free from projections. In a barn system, the top portion of the wall may be left open, or perhaps have a metal grille fitted so that the horses can see each other, but the lower part of the wall should always be of solid construction for safety. The use of partition grilles would depend on the type of horse being stabled and may not be suitable for mares with foals at foot who can be very protective. Grilles of this type are generally unsuitable for use with stallions; however, on some Thoroughbred studs a teaser may be stabled in the mare barn with such a partition grille to allow the mares to see him.

Floor construction is another important consideration. Proper drainage is essential and the location of some stabling may mean that additional artificial drainage needs to be laid in the foundation of the building. Concrete is one of the most common flooring surfaces used and it is deliberately left rough so that it is not slippery and, as the floor is laid, it will be made to slope to a specific point to encourage quick drainage. Concrete can be cold and will require sufficient bedding so that the horses can lie down without discomfort or injury. Soil or clay floors are more traditional but are difficult to keep clean and may pose an infection risk on commercial studs; this type of floor also tends to become uneven with wear, developing potholes and damp areas. Rubber flooring has become more common in recent years and some is designed to be used without bedding, being cleaned by sweeping out and washing down each day. However, rubber flooring is expensive to install, although savings can be made on the reduction in bedding and labour costs. Rubber is often used in conjunction with straw bedding for foaling boxes to provide a safe, warm surface when the mare foals.

Bedding in stud stables should be the best quality available, whatever type is used. It must be mould free, dust free and absorbent, but also economical, easy to use and dispose of. Straw is the most common bedding used with horses. Wood shavings are also popular but are generally unsuitable for foaling boxes as the shavings tend to stick to wet newborn foals and can get caught in their eyes.

Ventilation

Ventilation should be close to the top of the list of considerations when designing or planning stabling for horses. The horse was never meant to spend hours in stables surrounded by dust-laden air and it is no wonder that many of our modern performance horses suffer from respiratory problems. The principle of ventilation is to remove stale air and replace it with fresh air. When the weather is warm, we think of ventilation in terms of bringing cooling air in, but in fact, even in cold weather some movement of air is essential, especially in barn stabling. Ventilation should always be designed so that it does not create draughts. Horses of any age can cope well with cold temperatures but not with draughts.

Stable ventilation is usually provided by the use of windows and vents, but sometimes also by artificial means such as extractor fans. Vents are often situated in the roof so that the warm, stale air will rise and be removed without creating a draught at horse level. Eye level windows are common in stabling so that the horse can look out and these windows also provide additional ventilation. All windows should be glazed with safety glass and have a strong safety cover on the inside of the stable. Windows can also provide additional light in the stable. Artificial ventilation may be necessary in barn structures, especially those that have been converted for horse use and this is usually in the form of extraction fan units.

Condensation in stabling can cause serious respiratory problems, as the damp, humid environment is ideal for fungal spore growth. Proper ventilation will reduce the risk of condensation forming, as moisture does not condense as easily in moving air. Insulation also reduces condensation as it keeps the inside of the wall warmer than the outside. Roof insulation is also important for stabling. In a carefully designed stable, there should never be any need to shut the horse's top door unless some other form of artificial ventilation can be used.

Lighting

Good stable lighting is important on studs, especially when it is necessary to use artificial lighting programmes to promote early cycling in mares (see Chapter 1). Making as much use of natural light is important as it is free, but in the UK bright sunny days seem to be at a premium so some form of artificial light must be provided, especially as a lot of the activity on a stud tends to happen at night. Any light fitting should be correctly wired by a qualified electrician and suitable for use in an outdoor environment. For safety, stable lighting can be either recessed into the ceiling or wall, or fitted with a metal safety cover. Good quality lighting systems should be set up outside the stables as well as inside, illuminating loading ramps, access roads for arriving mares, etc., as well as for security purposes.

Veterinary Examination Facilities

Most commercial studs have specialized areas set aside for veterinary examination, but on smaller studs that use only traditional methods of breeding one examination area may be adequate for routine use (Figure 8.2)

On a large stud, the examination area may be in the form of an extra stable at the end of the barn/block, with a set of veterinary stocks for ease of use. These areas benefit from having electricity points, sinks, hot and cold water and work surfaces for equipment, etc. The positioning of spotlights to the rear of the stocks can also be very useful for the veterinary surgeon who will be performing mainly genital examinations on the mares. If the examination area is to cater for

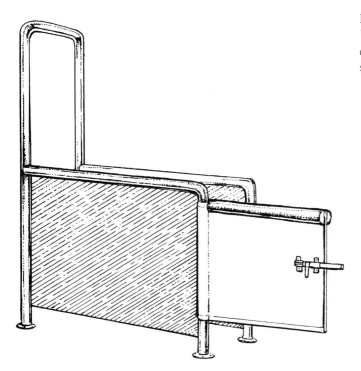

Figure 8.2
Veterinary examination stocks

mares with foals at foot, an additional enclosure at the head of the stocks, usually padded, is necessary. Examination areas, if enclosed, need to have two access doors to enable the mares to be safely and quickly moved through following examination. Some young mares may be nervous of entering the examination stocks and are calmed by having the exit door open as they arrive. The flooring used for examination areas must be non-slip and easily cleaned, for this reason rubber surfaces are popular. Studs that specialize in AI or ET may have complex examination and laboratory areas.

Storage

The main storage of forage and feedstuffs should be as far away from the stabling as possible to reduce the fire risk. All stored feed should be kept in rodent and waterproof areas. On large studs where lorry delivery of feeds occurs regularly, the access to the storage barn should permit the lorry to drive right up to the barn. Hay and straw is often stored in large quantities away from the stabling, with perhaps only a week's supply at a time being moved to areas next to the stabling. On smaller studs, lack of space may mean that large quantities of forage and bedding have to be stored close to the stabling, so all sensible fire precautions must be taken.

Although on many studs there may be little need for ridden work with the resident horses, a tack room is still necessary. Tack rooms situated close to the stabling provide storage for grooming kits, headcollars, bridles, etc., as well as often being used by the staff for tea breaks. The tack rooms should contain the first-aid kits for both horses and humans, hot and cold water, electricity points and, on larger studs, a telephone.

Machinery and fencing storage areas are generally hidden away on the stud, as they can be unsightly. Expert advice should be sought if there is a need to store large quantities of fuel, fertilizer and chemicals as many of these substances require specialist facilities.

The Breeding Area

Areas set aside for in-hand breeding vary in both size and quality depending on the number of mares bred each year and the breed of horse. On Thoroughbred studs, specially designed palatial breeding barns are commonplace. The stallions are the heart and soul of the large commercial studs and the breeding barn often reflects their importance. However, the most important design factor for any breeding area, whether it be a purpose built barn or the corner of a field, is that it must be safe for horses and handlers.

Any area used for breeding should have a non-slip surface which also provides a soft landing, as falls are not uncommon. For indoor areas, rubber flooring is common as it can also be easily cleaned. Wood chip is also popular but this can become very dusty with age and, when newly laid, can be slippery. Often, areas set aside for breeding need to double as exercise areas for the stallions, for example for lungeing, so this may need to be taken into account when deciding on the final design and surface. As was mentioned in Chapter 3, a mound or pit is often incorporated into the surface design to aid with the size difference between mare and stallion. The area used should be as dust free as possible to reduce the risk of airborne infection to the mare's or stallion's reproductive organs.

Most breeding areas are indoor and enclosed. This not only protects the horses and handlers from the elements, but also provides a safer environment, free from too many distractions. Large commercial studs with high numbers of visiting mares may have to work round the clock with their stallions to keep up with peak demand and to avoid missing any mares, so breeding areas that are used at night will need good levels of lighting.

The breeding barn often has two entrances, one for the mare and the other for the stallion. There may also be a teasing board set up within the barn, or just outside to permit the teasing of the mare immediately before mating. If the teasing board is located in the barn, it should be designed so that it can be folded back out of the way during mating.

For studs receiving a large number of visiting mares, loading ramps for horseboxes are commonplace, especially for those studs taking a large number of 'walking-in' mares (non-boarding mares who arrive just to be mated and are then taken back to their home stud). To reduce the risk of accidental transfer of infection, preparation areas for the mares to be mated are often separate from those used by the main stud. These preparation areas are similar to the veterinary examination areas and equipped with hot and cold water, electricity, facilities for mares with foal at foot, spare tack, etc. It is here that the mare will be checked against her paperwork and be prepared for mating.

The location of the breeding area in relation to the stallion stabling should also be a consideration. On some studs, the breeding barn is literally an extension of the stallion stabling, but the location of the barn will probably depend on the number of mares to be mated – leading an excitable valuable stallion a long distance in the dark to a breeding area can be nerve racking to even the most experienced handler. For studs specialising in artificial means of breeding, the breeding barn will need to be as close to the laboratory area as possible to reduce the risk of the sample being damaged by environmental factors, or the time taken before it can be properly prepared for storage.

The Foaling Box

Stabling set aside for foaling probably requires the most specialist facilities. Again, dependent on the size of the stud, the commercial value of the horses and the number of mares to be foaled, the foaling facilities may be just one or two stables, or perhaps an extensive foaling unit. Foaling boxes need to be larger than normal stables – often 4.5 m x 4.5 m – and permit constant unobtrusive supervision by attendants.

The construction of the foaling box is the same as that of a normal stable, but extra consideration needs to be given to safety. The floor must be non-slip and be easily cleaned between each foaling. Ideally, the walls and ceiling should be sealed so that the entirety of the inner surface can be pressure washed and disinfected between each mare. Any stable furniture such as water and feed buckets should be kept to an absolute minimum, or boxed in to the corner of the stable so that there is no risk of injury. Foaling boxes commonly have heat lamps permanently installed, particularly on those Thoroughbred studs where the mares begin to foal in early January. All heat lamps should be of the type designed for use in stables and sited well out of reach of the mare.

Ideally, the foaling box should be designed with rounded corners and with two entrances; a normal entrance at the front of the stable for the horses, and one at the rear for attendant/veterinary assistance, with a peephole or hatch inset into the door for ease of supervision. The stable doorways should be wide enough for a heavily pregnant mare to fit through easily.

Sitting up rooms attached to the foaling boxes are common. These rooms often have a small window or hatch that permits the attendant to quietly observe the mare/mares during the night. Nightwatchmen generally remain with the pregnant mares for the whole of the night, so some comforts such as an armchair, television and kettle, etc., are necessary. It is important that the sitting up room is kept clean and tidy at all times to reduce any infection risk to the horses and to ensure that the foaling equipment stored there remains as clean as possible. Smaller studs may monitor their mares by coming and going from the main house, but it is most important that great care should be taken to keep disturbance to a minimum. Monitoring equipment, such as closed-circuit cameras, are now commonplace on studs and the foaling box should have the facility to easily install a camera for the duration of the breeding season. If closed-circuit cameras are used, fitting infra-red lights can also be beneficial as this enables the mare to be seen clearly with the minimum of disturbance to her.

Paddocks and Fencing

Well-managed paddocks are an essential part of any stud (Figure 8.3). If there is sufficient good quality land available, it can significantly reduce day-to-day costs such as labour and supplementary feed. Horses that can be turned out each day, even for just a short while, do not tend to suffer from the same level of respiratory problems and boredom related stable vices.

A minimum of approximately 0.5–1 ha of good quality grazing is recommended per horse, but more than this is often necessary in areas when the land and grazing quality is poor.

As the age range of horses varies considerably on the majority of studs, it is common for the paddocks to be set out in varying sizes. It is common for the land close to the stud buildings to be divided up into small paddocks, that is, for use by mares with young foals at foot; this type of paddock is called a nursery paddock. Stallion paddocks are also often sited close to the stallions' stabling, on large studs each stallion may have a paddock leading directly off from his stable. The remainder of the land located further way from the stabling is often divided up into large areas for use by the barren or maiden mares or by groups of older youngstock, etc.

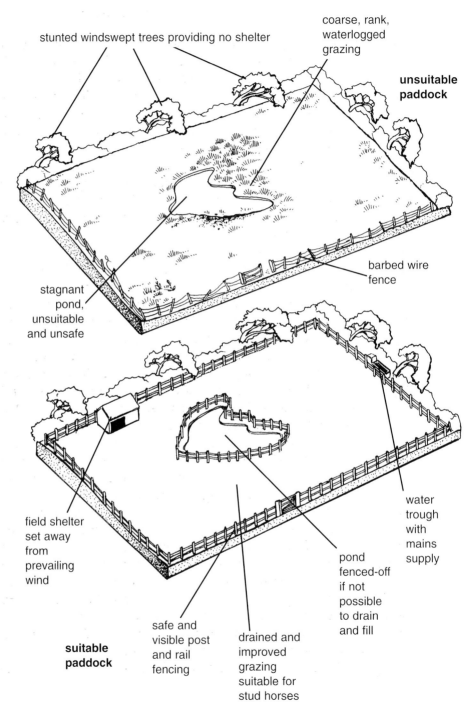

stunted windswept trees providing no shelter

coarse, rank, waterlogged grazing

unsuitable paddock

stagnant pond, unsuitable and unsafe

barbed wire fence

field shelter set away from prevailing wind

water trough with mains supply

pond fenced-off if not possible to drain and fill

safe and visible post and rail fencing

drained and improved grazing suitable for stud horses

suitable paddock

Figure 8.3
Before and after examples of a paddock adapted for stud use: (top) unsuitable paddock; and (bottom) suitable paddock

When planning the paddock layout for a stud, it is wise to think of what horses will need throughout the whole of the stud year. This means that the most efficient use can be made of the land all year round without having to overgraze some areas whilst others go to waste because they are too far away from the buildings or too large for regular use. If there is sufficient land, some studs may also choose to grow grass crops for hay, etc., to make best use of the larger isolated paddocks.

The needs of a grass kept horse are relatively simple – aside from grass, they need water and shelter. As we all know, water is one of the most important aspects of a horse's diet and there must be free access to clean fresh water at all times. Automatic water troughs are the most common way of supplying water to grass kept horses. These troughs have water piped to them, either from mains or spring/bore supplies and operate on a ballcock float so that they do not overflow. An important point to remember regarding automatic drinkers is that the water pressure should be enough to allow the refill time to be as quick as possible. It may seem unimportant but in paddocks used for lactating mares, there is a constant demand for water especially in the warmer months. Water troughs should be set into the fence line if at all possible to reduce the risk of accidental injury and they should be easy to empty and clean. The use of rivers or streams for watering stud horses is not recommended in most cases.

Shelter can be provided naturally by the geographical position of the paddock, or by artificial means such as planting trees to form a shelterbelt, or by the use of field shelters. Tree or hedge shelterbelts are excellent for stud use, but they can take many years' growth to be effective. Field shelters are generally constructed to have three sides and a roof. Ideally, the open side should face south-westerly in the UK but, if not, away from the prevailing wind. This allows for the maximum shelter from winter winds and also direct sunlight in summer. If field shelters are to be used, they must provide sufficient space for each horse, especially if the horses are to be fed in them as well. Field shelters are not commonly used for mares with foals at foot due to increased risk of injury through fighting. The inside of the shelter should be constructed following the same principles as that of a stable and be free from projections and fully boarded. Ideally, the corners should be rounded to avoid any horse becoming trapped in the corner by a more dominant paddock companion.

If the horses are to receive concentrate feeds out at grass, sufficient well spaced out feed trough/buckets should be provided to allow all the horses to feed at the same time. Some studs will put out one more bucket than there are horses to reduce the risk of the slower eating horses missing a feed. If creep feed is to be provided in the paddock for foals then specially constructed feed pens are often used. The use of feed bins fixed to the paddock fence is not recommended in paddocks when there is a large number of horses or with youngstock. A safer alternative is to use buckets set out on the ground, ideally shallow buckets set into a tyre to avoid spillage and injury – these are very useful with youngstock who tend to get more boisterous than adult horses at feed time. If the land is dry enough, some studs will provide piles of feed for the horses directly onto the grass rather than in buckets.

Fencing for stud use should be strong, easy to maintain, highly visible and safe (Figures 8.4 and 8.5). Whatever type is chosen, it is important that it is checked regularly and any repair made immediately. Fencing for stallion paddocks should be much higher than that for normal horse use. In addition, fencing used for mares with young foals should have a low bottom rail to reduce the risk of the foal rolling underneath. Wherever possible, corners should be eliminated from all stud paddocks by using curving fence lines and/or boarding out the corners.

Wood fencing, such as post and rail, is perhaps the most popular for stud horses. It is attractive and very visible. Unfortunately, it is initially expensive to install, does require regular maintenance and is susceptible to damage from horses that chew or crib. Wooden fencing is often pressure treated with preservatives prior to installation, but should be repainted with a suitable non-toxic preservative at least every two years to help increase its life span. Post and rail fencing may be further enhanced by planting a hedge alongside; this not only improves the appearance of the fence but also provides further strength and shelter.

Specialized wire mesh fences, constructed for horse use may also be combined with wood post and rails. The mesh is designed to form

Figure 8.4
Three examples of materials used for enclosures

Post and rail fencing

Post and rail with a hedge

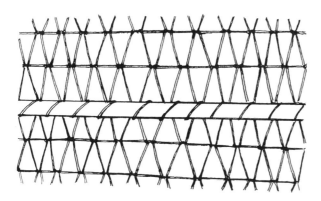

Wire mesh fencing material, specifically designed for use with horses

Figure 8.5 *Youngstock grazing in a well-fenced paddock. This type of fencing is also suitable for mare and foal paddocks as the rails are closely spaced for safety and the bottom rail is close to the ground to reduce the risk of a foal rolling underneath*

small V shapes and this provides strength without any risk of a horse catching a foot through the mesh. The mesh also stops other animals such as dogs and rabbits from being able to easily get in to the paddock. Wire mesh of this type is almost maintenance free, but must have some form of rail support structure for strength and to increase its visibility.

Electric fencing is generally not used on its own as most types are designed more for cattle or sheep. It can be a useful form of fencing for horses when combined with post and rail; often an electrified wire is run along the top rail to reduce the amount of wood chewing or cribbing. Stallion paddocks are often further secured with an electrified wire, held some 13–15 cm away from the main fence line by special supports to help keep the stallion well away from the fence. This can be of particular use with a stallions that may be very territorial and aggressive when other horses are led past. Occasionally, electrified tape is used on studs for restricting horses to one part of a large field usually to reduce the amount of grazing available at any one time. This form of temporary fencing is generally unsuitable for Thoroughbred youngstock, stallions and for mares with foals at foot, but may work well with quieter adult Thoroughbreds and native breeds. Any electrified fencing needs to be clearly marked and checked daily. Wire fencing of any type can be dangerous to horses if not properly secured and kept taut. Proper installation is therefore essential for any electrified fencing. The power supply to the fencing is often provided by a portable fencing battery, which has an audible intermittent click when turned on.

Barbed wire is one of the most dangerous types of fencing. It is totally unsuitable for use on a stud as horses can easily become entangled in it and will tend to panic, trapping and injuring themselves yet further.

Fire Safety and Emergency Procedures

Reducing fire risks is one of the most important aspects of stud design. Prevention is always better than a cure and never more so than in the case of a fire outbreak on a stud. Free advice can always be sought from the local fire station as to safe practice and emergency procedures, or specialist equine advice can be obtained from advisory bodies such as ADAS (Agricultural Development Advisory Service).

When designing a stud, consideration should be given to the location of storage barns so that they are away from the main heart of the stud. Storage areas tend to be large and those used for hay/straw storage can present a high fire risk. Some studs will construct several smaller forage barns to reduce the risk of a major fire, as well as having several smaller blocks of horse stabling to reduce the risk of all the horses being lost should a fire occur. This is obviously not going to be practical or possible for every stud, but it is a consideration.

All buildings on a stud should be constructed of a fire resistant material, or be treated so that it is slow to spread fire. However, despite treatment with resistant substances, most stud buildings are quick to spread fire because of the very nature of what is kept in them, straw, hay, bags of concentrate feeds as well as dust are all very combustible. Hay is of particular concern as it can spontaneously combust if packed into a barn when still wet due to a fermentation process that can occur. All storage areas and walkways should be kept as clean as possible and all rubbish removed immediately.

Electrical wiring to light fittings, heaters, etc., is another major source of fire on studs. There are now safety regulations concerning how often wiring needs to be checked by a competent professional; all new installations should be carried out by a qualified electrician.

Other safety considerations are fire notices prominently placed for location of fire extinguishers, no smoking signs which are enforced, safety mesh used to cover all light fittings and all fittings used to be of the type safe for outdoor use. Heaters and heat lamps are commonly used on studs; to reduce fire risk, these should be of the type that is permanently installed to reduce the risk of unreported damage or of unsafe temporary installation. Lightning conductors are another consideration – lightning strike is reported to cause a high number of barn fires in the UK each year.

High levels of stable management will also dramatically reduce fire risk. Keeping barns swept and removing the build up of dust and cobwebs from the rafters are good practice. All stud staff should understand safe practice with regard to using any inflammable materials or chemicals and only those qualified to use such substances should do so.

If a fire should break out on a stud, early detection is important to reduce the risk of loss of life to both horses and staff. There are many fire alarm systems available for studs – they should be easy to identify, reliable and should be able to detect fire in as many parts of the stud as possible. Water or chemical sprinklers may be installed on larger studs, as may fire doors to reduce the risk of spread from one side of the barn to another (some forms of chemical sprinklers are unsuitable for use in stabling as they can be toxic – advice should be sought from a fire officer and/or veterinary surgeon prior to installation). Portable fire extinguishers form the majority of any equipment kept for fire purposes and, for commercial studs, they are a legal requirement.

Each type of extinguisher is designed for a certain type of fire: that is, water cylinders for hay barns and dry powder for workshops and offices. Expert advice should always be sought before fire extinguishers are installed to ensure the correct ones are purchased and all extinguishers should be serviced every year, regardless of whether they have been used or not. The location of each fire extinguisher should be clearly marked and all staff should be aware of how to use them in an emergency.

Each stud should have a carefully planned emergency procedure to follow in the event that fire should break out. Everyone who works on the stud should be familiar with the procedure and, if necessary, practice fire drills can be carried out.

Any emergency procedure will be tailored to fit the individual stud: what works for the large commercial stud is impractical for the small private stud and vice versa. However, the basic principles remain the same.

- On detection, all people in the immediate area should be evacuated, the fire services should be called and the general alarm raised.
- Small fires may be contained by prompt action with an extinguisher, but no personal risk should be taken as stable fires spread quickly.
- All stud entrances should be kept clear for the arrival of the fire services and a staff member should be at the entrance to direct the fire service.
- Horses should be evacuated if possible, but without risk to human life. A carefully planned fire emergency procedure means that at this stage, the staff know where to take the horses and can act in a calm manner. Ideally, headcollars and lead

ropes should be kept by each horse's door in the event of fire breaking out. Blindfolding frightened horses can encourage them to move from a burning stable but, again, not if there is any risk to human life by doing so. Stables should never be locked in case a fire should occur.

- Horses should be led to safe areas – ideally the paddocks to be used in a fire will have formed part of the fire drill and each member of staff will know which horse goes where. Great care should be taken in evacuating stallions to a suitable and safe location.
- All safety procedures should be reviewed regularly as it can and does save lives.

Studfarm Records

Detailed records are essential to any well-managed stud, regardless of its size. Commercial studs are required to keep detailed accounts and payroll files alongside management records, whereas small private non-commercial studs may have need of only the basic management accounts. Each stud devises its own record system according to what works well for that situation but the basic principles of record keeping are the same throughout; for many breeders, commercial or private, time is at a premium and so any record system should be as efficient to use as possible.

Computer programs have superseded many of the older paper based records and many bespoke programs allow daily records to run alongside commercial accounts which reduces dramatically the time taken and number of people needed to manage the record system. On high-level commercial studs, laptop computers are used by stud staff which allows immediate access to all the individual horse records and the direct update of a horse's record, for example, during the daily veterinary round.

In this section we briefly outline the main records kept by studs; however, those records required by law, such as accounts and VAT, are not included as they are beyond the scope of this book. Whichever method is used, the basic requirements of any system are that all records should be easy to find, easy to understand, and accurate and simple to use.

Identification

The requirement for accurate identification of each individual horse is vital for all studfarms. All breed societies require horses to be accurately identified before they can be registered and most studs will insist on viewing the horse's identification and registration papers before any mating occurs. The Thoroughbred breed has, to date, been the most accurately identified due to the complex registration procedure required for inclusion in the General Stud Book. Sadly, for many other breeds, the level of accurate identification required for registration purposes varies enormously. From a stud manager's point of view, it is imperative that each horse has a clear identification record available on arrival so that there can be no doubt as to the identity of the individual horse. On arrival at stud, the horse's paperwork will be checked and identification verified immediately. Each horse must have an identification tag attached to its headcollar, if it does not already have one – some studs use a number system in addition to the horse's name and/or pedigree.

By 2003, there will be a requirement for every horse in the UK to have a passport, which clearly identifies the individual. This is new legislation and is in response to the devastating outbreak of foot and mouth disease during 2001. In the past only some breed societies have issued passports and many of these are not as accurate as one might hope. The Thoroughbred breed, which has had a passport system for some time (Figure 8.6), has been blood typed for identification purposes for several years and, more recently, all Thoroughbreds are also DNA tested and microchipped prior to registration. This ensures that accurate identification and confirmation of parentage is possible prior to registration. The new legislation covering all horses is good news for the whole horse population in every respect.

Passport or identification papers presently used in the UK have a

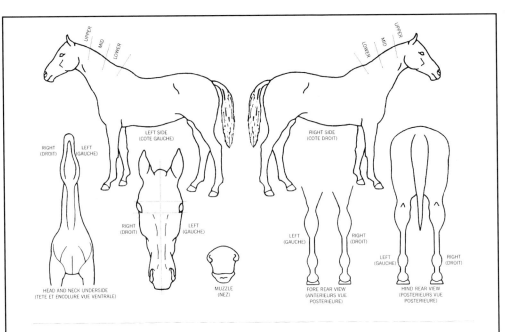

RIGHT (DROIT) — LEFT (GAUCHE)	LEFT SIDE (COTE GAUCHE)		RIGHT SIDE (COTE DROIT)	
HEAD AND NECK UNDERSIDE (TETE ET ENCOLURE VUE VENTRALE)	RIGHT (DROIT) — LEFT (GAUCHE)	MUZZLE (NEZ)	LEFT (GAUCHE) — RIGHT (DROIT) FORE REAR VIEW (ANTERIEURS VUE POSTERIEURE)	LEFT (GAUCHE) — RIGHT (DROIT) HIND REAR VIEW (POSTERIEURS VUE POSTERIEURE)

VETERINARY CERTIFICATE OF AGE AND MARKINGS FOR FOAL REGISTRATION

Please read instructions overleaf before completing this form.
Once completed give this form to the owner/agent.

* THESE ITEMS ARE BASED ON INFORMATION SUPPLIED BY THE OWNER OR THEIR AGENT

COLOUR (ROBE)	SEX (SEXE)	*DATE OF BIRTH (ANNEE)	*SIRE (PERE)	*DAM (MERE)
		/ /		

HEAD (TETE)	
NECK (ENCOLURE)	
LIMBS (JAMBES) — L.F. (A.G.)	
R.F. (A.D.)	
L.H. (P.G.)	
R.H. (P.D.)	
BODY (CORPS)	
ACQUIRED (MARQUES ACQUISES)	

Please affix Weatherbys microchip barcode here.

NAME AND ADDRESS OF VETERINARY SURGEON (IN BLOCK CAPITALS)

I certify that I have read and understood the instructions overleaf. I have been given the pedigree details by the owner/agent who has assured me that they have confirmed the identity of the dam against her passport. I have also,
*a) bloodsampled the foal,
*b) inserted a Weatherbys Microchip into the foal,
*c) scanned and read a Weatherbys microchip previously inserted.

Date of examination
/ /

Signature of Veterinary Surgeon
(not to be the breeder, owner or trainer of the horse)

5795331801

* Please delete as appropriate

Figure 8.6 (opposite page)
An example of a Thoroughbred identification card which must be completed by a veterinary surgeon for registration purposes. (Reproduced by kind permission of Weatherbys Group Ltd.)

diagram of a horse so that each individual animal's markings and hair patterns can be shown. In other countries is it common for some breed societies to also request drawings or photographs of the horse's chestnuts to be included on the identification page. (Chestnuts are the horny lumps of tissue found on the inside of the horse's legs and they were thought to be unique for each horse, rather similar to that of a fingerprint in humans.) Passports often include details of each horse's ownership, vaccinations record, import/export and, in the case of Thoroughbreds, details of checks carried out at racecourses.

Microchipping of animals is relatively commonplace in the small animal world. It provides an accurate form of identification/ownership if a microchip scanner is available. The chip, which is about the size of a grain of rice, is inserted into the neck muscle of the horse, along the crest and contains a bar code number that is unique to that horse; it can be easily read using a special scanner. The chip should remain in place for the whole of the horse's lifetime and, in the future, it may be possible to add further information to the chip for management purposes. Other forms of accurate identification are freeze branding and tattooing (not common in the UK).

Breeding Records

The quality and complexity of breeding records kept varies from stud to stud. Essentially, each broodmare should have a detailed breeding record and, ideally, a copy should go with her when she travels to a new stud. In Chapter 3, we discussed the importance of having an accurate breeding history for each mare and, in many cases, as mares tend to lead a nomadic lifestyle from stud to stud depending on the stallion they are to mated by, these records form the only detailed information available on an individual.

Breeding records should include the daily teasing record, veterinary treatment, foaling details, mating details, routine treatments such as farrier and worming, as well as any movements to another part of the stud. Keeping a record of movements on the stud may seem unnecessary in the case of a small breeding operation, but can prove vital in the case of a disease outbreak where it is important to know which horses have been in contact. Commercial studs, too, may have the management records linked to the accounts system to allow accurate billing for keep fees, etc.

Stallions on the stud also need detailed and accurate records. Most stallions will have a record sheet detailing routine matters such as worming dates, etc., but they should also have diary records of which mares they covered, how many mares were covered each day and results of semen evaluations taken (especially with those stallions used for artificial conception methods). A depth of detail recorded for each stallion during the breeding season can provide very useful information for the breeder/stud manager to ensure that the best fertility rates, etc., are maintained. When sperm production was discussed earlier, we saw that a drop in fertility may be reflecting a problem that happened some weeks previously; by keeping detailed records of mating and pregnancy rates for each stallion, it may be possible to identify a problem before it occurs in the future.

Nomination contracts are an essential record for commercial studs standing stallions – the form of the contract varies and they are most commonly used on Thoroughbred studs. A nomination contract is issued when a mare is booked to be mated by a stallion and it details the chosen stallion and the nominated mare, as well as the agreed stud fee. The form is a legal contract which is signed by the stallion owner or bloodstock agent as well as by the mare owner prior to the mare arriving at the stallion stud.

For Thoroughbreds, several different types of nomination contract are used dependent on the terms of payment for the stud fee. The most common ones are: 1 October terms when the stud fee is payable in full on 1 October following the breeding season unless the mare is certified barren by that date; live foal terms when the stud fee is payable in full if the foal is born alive (the contract may specify that the fee is payable if the foal is born alive and lives for at least 24 or 48 hours, sometimes seven days, after birth); split fee terms which means that part, usually half, of the fee is payable immediately and the remainder on 1 October.

The type of nomination contract issued is often dependent on the popularity of the stallion and value of his stud fee. Those stallions standing in the top price bracket are rarely available on live foal terms and live foal contracts are generally considerably more expensive than those issued on 1 October terms. For very popular stallions, contracts are often only issued to high-quality mares of the very best pedigrees.

Records kept for youngstock will depend on the individual stud. It is common for the foals born each year to have their details kept alongside their mother's until they are weaned when they will then have their own files. Keeping youngstock records in as much detail as that for the other stud horses can be very helpful and some studs record growth and development for each youngster in great depth. Identification records for foals should be completed as soon as possible each year and most foals require to be registered with the relevant breed society by the time they are six months of age. (Horses can usually be registered at any time in the first few years of their lives, but the breed society charge often increases with age to encourage owners to register early.)

For many commercial studs, keeping accurate records not only helps increase the efficiency of management on the stud, but is also invaluable in the case of legal action. Sadly, in this day and age, litigation has become almost standard practice and many breeders are sued at the proverbial 'drop of a hat', often for something that turns out to be a complete misunderstanding. Any legal action is costly in terms of money, but more so in terms of reputation especially if the situation could have been avoided.

Index

abdomen 4, 17, 18, 76, 84, 128

abnormal behaviour 36–7, 130

abortion 44–5, 47, 48, 77, 81, 85, 86–101

accessory glands 18

adaptive period 127, 130

administering drugs 23, 66, 83, 152

AI (artificial insemination) 12, 21, 24, 25, 49, 50, 51–4, 56, 99

allanto-chorion 74, 110, 121, 122

Allen, Professor 'Twink' xi, 56, 70

amnion 74, 110, 114, 115, 121

ampulla 18

anatomy of the mare's reproductive system 1–14

anatomy of the stallion's reproductive system 15–25

Animal Health Trust 70

anoestrus 8, 10, 13, 36

anus 1, 3, 41, 109, 125

artificial conception 29, 54, 56, 182

artificial lighting 10–11, 164

aspects of fertility 14, 15, 17, 18, 21, 22, 23, 24–5, 28, 39, 43, 51, 54, 55, 56, 59, 124, 182

AV (artificial vagina) 49, 51, 52

barren mare 27, 35, 82, 160, 182

bladder 2, 3, 5, 16, 18, 74, 128

blood 5, 7, 8, 12, 19, 20, 47, 73, 74, 79, 95, 117, 125, 129

boots 59, 61, 65, 105

breed types 9, 49, 57, 62

breeding area 54, 57, 58, 59, 62, 66, 163–8

breeding methods 10, 11–12, 20, 22, 26–66, 69, 77, 82, 89–90, 92, 132, 167, 181–3

breeding rolls 62–3

broad ligaments 4–5, 6

bulbourethral gland 16, 18

Caesarean section 124

cape 60, 61

Caslick operation 41, 109

CEM (contagious equine metritis) 44, 45, 46, 138

cervix 2, 3–4, 19, 40, 43, 45, 53, 62, 63, 77, 85, 87, 108, 122

choosing a mare 26–9, 68, 137

choosing a stallion 29–31, 68

cilia 6

CL (corpus luteum, or yellow body) 9, 13, 75

clitoral sinuses 2, 40

clitoris 1, 2, 36

Codes of Practice 43, 44, 47, 48

coital exanthema 47, 48

colostrum 85, 104, 106, 120, 129, 130, 133, 135

conceptus 40, 50, 67, 72, 75, 78, 79, 86, 92
cryptorchid (or 'rig') 17
deformities 88, 89, 128, 130–2
diarrhoea 84, 134
dioestrus 7, 8, 9, 10, 13, 32, 36, 77
DNA testing 70, 179
E. Coli 96
early handling of the foal 143–6
eCG (equine chorionic gonadatrophin) 75, 76, 79
egg (or ovum) 6, 8, 20, 21, 22, 55, 56, 57, 67, 72, 84
EHV (equine herpes virus) 43, 44, 47, 82, 83, 96–8, 133
ejaculation 18, 19, 21, 53, 65
endometrial cups 75
endometritis 39, 43, 72, 93
endometrium 4, 40, 44
epididymis 18
Equine Fertility Unit 56, 69
equine influenza 43, 48
ET (embryo transfer) 12, 49, 50, 55–7
EVA (equine viral arteritis) 44, 45, 47, 98–100, 138 .
evolution of the horse 9–10, 32
eye 7, 8, 10, 99, 132, 163
faecal contamination 3, 41, 116
fallopian tubes 1, 5–6, 20–1, 57, 72
farrier attention 83, 132, 143, 150, 182
fertilization 6, 20–1, 54, 55, 56, 67, 70, 72
fire safety and emergency procedures 175–8
foal exercise and behaviour 146–7
foal heat 38, 134
foal proud 66
foaling 102–42
foaling box 98, 103, 107, 112, 163, 168–9
foaling equipment 104–5, 143–6
foaling problems 121–7
foetal development 72–5
foetus 4, 39, 67, 73, 74, 76, 79, 80, 82, 86, 87, 89, 92, 93, 94, 95, 97, 102, 109, 129

follicles 6, 7, 8, 9, 13, 37, 75, 76
foot and mouth disease 44, 179
fostering 99, 135–42
FSH (follicle-stimulating hormone) 8, 9, 13, 20
genetic research 56, 69, 70, 89, 94
genotype 69
gestation 8, 67, 70, 79, 86, 87, 89, 92, 94, 95, 102–9, 125
GnRH (gonadatrophin-releasing hormone) 8, 20
Grimwade, Joe x, xi
haemolytic disease 129
haemorrhage 84, 125–7
hernia 17, 132–3
herpes 44, 47, 48, 96–7
hippomane 74–5
hobbles 62
hormonal activity 4, 6, 7, 8, 9, 12, 13, 14, 20, 38, 39, 75–6, 79, 90, 92–3, 102
Horserace Betting Levy Board 43, 46
hymen 3
hypothalamus 1, 7, 8, 20
identical twins 56, 70, 71
IgG levels 106, 120, 121, 133
in-hand breeding 23, 49, 54–5, 64, 166
infection 2, 4, 5, 28, 30, 33, 39, 40, 43–9, 51, 53, 58, 80, 81, 82, 84, 85, 93, 95–100, 133, 134, 135, 138, 156, 160
infundibulum 5
inguinal canal 17
injury 12, 17, 22–3, 30, 32, 34, 43, 50, 51, 59–61, 83, 134, 139, 171, 172
insurance 101
interstitial cells 20
isolation procedures 7, 44, 48, 49, 88, 97, 98, 100–1, 138
Jockey Club ruling 80
Kentucky research 89
kidneys 2, 16, 17
Klebsiella pneumoniae 44
labia 1, 3

lactation 13, 82, 148

leg strap 62

letting down 12

LH (luteinising hormone) 8, 20

lymphatic system 12

maiden mare 3, 27, 28, 35, 41, 59, 61, 62

malpresentation 124

mare 1–14, 26–9, 55–7, 58, 59–63, 66, 67–72, 76–85, 92–4, 102–35, 144, 166, 168, 181

Mare Reproductive Loss Syndrome 89

mare's genital tract 1, 2, 3, 5, 8, 19, 39, 40, 41, 95, 108, 124, 125

mating 3, 8, 9, 20, 21, 24, 30, 32, 33, 36, 38, 39, 45, 47, 50, 54, 55, 57–66

meconium 116, 128,

monitoring 102–4

neck rope 61–2, 65

nerve cells 4, 5, 17

newborn foal 83–142

NMS (neonatal maladjustment syndrome) 130

nutrition 81–2, 147–9

obese mare 11

oestrogen 8, 75, 79

oestrous cycle 1, 7–14, 32, 38–9, 56, 75, 92

oestrus 2, 3, 7, 8, 9, 10, 12, 13, 20, 27, 32, 35, 36, 37, 38, 45, 55, 63, 66, 70, 77, 86, 92

orphan foal 135–42

ovaries 1, 2, 5, 6, 8, 9, 13, 22, 37, 38, 75, 78

ovulation 6, 7, 8, 10, 38, 69, 75, 77

paddock (or free) breeding 49, 50–1

paddock fencing 33, 169–75

pelvis 84, 106, 110

penis 2, 3, 15, 18, 19, 23, 40, 47, 59, 62, 65

performance horse 11, 55, 81, 163

phantom (or dummy) mare 51–2

phenotype 69,

pheromones 36

physical fitness of mare 11, 12

physiology of the stallion 19–20

pineal gland 1, 7, 8

pituitary gland 1, 7, 8, 9, 13, 20

placenta 4, 56, 67, 68, 72–4, 75, 90, 91, 93–5, 96, 97, 110, 115, 117, 118, 119, 120, 122, 129

PMSG (pregnant mare's serum gondatrophin) 75, 79

pneumovagina 41

polyoestrous breeder 8

Pouret's operation 41

pregnant mare 8, 36, 38, 44, 48, 67–85, 103, 104, 109, 152, 168

premature foal 76, 86

preparation for foaling 105–8

problem mare 14, 39–43

problems affecting newborn foals 127–35

progestagen 76

progesterone 4, 9, 38, 75, 76, 79, 92, 93

prostaglandin 9, 75

prostate gland 16, 18

proven mare 27, 28

proven stallion 29

Pseudomonas aeruginosa 44

racehorse 11, 12, 80, 81

rectal palpation 7, 37, 77, 124

rectovaginal fistula 125, 126,

rectum 2, 7, 37, 78, 125

Rossdale, Dr Peter xi, 13

routine care in pregnancy 83, 84

Royal Studs 70–1

safety 15, 33, 35, 57–8, 108, 127, 136, 139, 146, 160, 162, 163, 164, 168, 175–8

Salmonella 96

scrotum 16, 17, 19

semen 3, 4, 16, 18, 19, 20–5, 45, 47, 51, 53, 66, 98, 99

seminal plasma 18, 21

seminal vesicles 18, 19

sheath (or prepuce) 2, 19, 23, 45, 99

shedder 47, 49

silent heat 13

smegma 2, 19, 45

sperm (spermatozoa), 6, 15, 17, 18, 19, 20, 21, 22–5, 53, 54, 65, 67, 72, 89, 182

spermatic cord 17, 18

stallion 2, 7, 8, 15–22, 26–31, 32, 33, 35, 37, 45, 47, 49, 50, 51, 52, 54, 55, 57–8, 63–6, 68, 82, 89, 98, 99, 160, 166, 167, 169, 172, 175, 182, 183

Staphylococci 96

sterile stallion 33

strangles 44, 48, 49

Streptococcus 96

stress in pregnancy 80, 90, 94, 95, 100, 121, 125, 144

Strongyloides westeri 83, 84, 151

stud design and administration 10, 11, 12, 13, 15, 25, 29, 30, 31, 32, 36, 43, 44–6, 48, 51, 158–83

stud fees 28, 29, 30, 82, 182, 183

stud hygiene 45, 47, 89, 98, 100, 133, 160

stud staff, duties of 14, 15, 32, 36, 57, 58, 62, 63, 64, 100, 102, 103, 104, 138, 146, 160, 176, 177, 178–83

swabbing 19, 39, 40, 43, 44, 45, 49, 85, 97, 99

syndication (or leasing) 31

teaser stallion 32–3

teasing 31–9, 59

teeth 149–50

testicles (or testes) 16, 17, 18, 19, 20

testosterone 17, 20

tetanus 43, 83, 151

Thoroughbred Breeders' Association 101

Thoroughbred horse 9, 10, 12, 15, 24, 27, 30, 31, 32, 38, 40, 43, 44, 45, 48, 50, 54, 55, 69, 70–1, 74, 82, 92, 94, 98, 103, 105, 108, 116, 120, 136, 149, 150, 158, 175, 179, 181, 182

transporting horses 100

trauma 43, 90, 95, 128–9, 130, 135, 155

twin pregnancies 56, 69, 70–1, 73, 77, 78, 79, 90–2

twitch 62

udder 85, 104, 105–6, 117, 120, 133, 136, 139, 141, 156, 157

ultrasound 38, 72, 77, 78, 79, 92

umbilical cord 74, 94, 117, 128, 132, 133

ureters 16

urethra 2, 3, 18, 19, 45, 65, 128

urine 3, 18, 36, 41, 74, 79, 128, 138

uterine haemorrhage 84, 120

uterus 1, 2, 3, 4, 5, 6, 9, 19, 21, 38, 39, 40, 44, 45, 50, 53, 57, 65, 67, 70, 72, 73, 75, 76, 77, 78, 79, 84, 90, 91, 93, 94, 95, 97, 110, 117, 119, 122, 124, 127, 129, 130, 132, 134

vaccination 43, 44, 47, 82–3, 90, 99, 100, 150–1, 181

vagina 1, 2, 3, 4, 5, 8, 9, 41, 43, 65, 66, 85, 92, 109, 113

vas deferens 18

venereal infection 2, 19, 33, 44, 45

vernal transition 10

vestibule 2, 3, 95

veterinary examination 7, 13, 17, 28, 36, 37, 38, 39, 43, 51, 53, 55, 57, 66, 72, 77, 78, 79, 82, 84, 85, 87, 89, 99, 105, 107, 120, 121, 124, 128, 129, 134, 150, 164–5, 182

vulva 1, 2, 3, 8, 19, 28, 39, 41, 42, 47, 96, 107, 109, 121, 125

washing 58, 65, 117, 162

weaning 143–57

Weatherbys Group Ltd 181

winking 36

worming 83, 151, 152, 182

youngstock 27, 81, 96, 143, 150, 159, 160, 172, 183